Exploring Spatial Analysis in Geographic Information Systems

Yue-Hong Chou

Exploring Spatial Analysis in Geographic Information Systems

By Yue-Hong Chou

Published by:

OnWord Press
2530 Camino Entrada
Santa Fe, NM 87505-4835 USA

First Edition, 1997

SAN 694-0269

10 9 8 7 6 5 4 3 2 1

Printed in the United States of America

Library of Congress Cataloging-in-Publication Data

Chou, Yue-Hong, 1952-
   Exploring spatial in geographic information systems /
Yue-Hong Chou.
         p.   cm.
   Includes index.
   ISBN 1-56690-118-9
   1. Geographic information systems. 2. Spatial analysis
(Statistics)   I. Title.
G70.212.C475   1996

910' .285—dc21   96-37702
                                           CIP

Trademarks

All terms mentioned in this book that are known to be trademarks or service marks have been appropriately capitalized. OnWord Press cannot attest to the accuracy of this information. Use of a term in this book should not be regarded as affecting the validity of any trademark or service mark.

Warning and Disclaimer

This book is designed to provide information on quantitative techniques for spatial analysis employed with geographic information systems. Every effort has been made to make the book as complete, accurate, and up to date as possible; however, no warranty or fitness is implied.

The information is provided on an "as is" basis. The authors and OnWord Press shall have neither liability nor responsibility to any person or entity with respect to any loss or damages in connection with or arising from the information contained in this book.

About the Author

Yue-Hong Chou, associate professor in the Department of Earth Sciences at the University of California-Riverside, specializes in geographical information systems, transportation modeling, location theory, and spatial statistics. He earned his Ph.D. in geography at Ohio State University. In addition to teaching and research, Yue-Hong has served as a consultant for the U.S. Department of the Navy, U.S. Forest Service, California Department of Fish and Game, California Department of Forestry and Fire Protection, County of Riverside, Southern California Association of Governments, California Air Resource Board, and the South Coast Air Quality Management District. His publications have appeared in numerous journals, including *Geographical Analysis, International Journal of Geographic Information Systems, Photogram-*

metric Engineering and Remote Sensing, Geographical Systems, Transportation Research, Transportation Planning and Technology, Journal of Transport Geography, Environmental Management, International Journal of Wildland Fire, Forest Science, and *Lecture Notes in Computer Science.*

Acknowledgments

This book would not have been possible without the generous support of Jack Dangermond, president of Environmental Systems Research Institute (ESRI), and a leader in the GIS industry. I am especially grateful to William Miller and Michael Phoenix of ESRI for their enthusiastic encouragement and constant assistance throughout the development of this book.

Barbara Kohl of High Mountain Press made tremendous editorial efforts and significantly improved the manuscript. Raymond J. Dezzani of Boston University, Jay Lee of Kent State University, and Gary Kabot, Randy Worch, and Steve Kopp of ESRI provided constructive comments on an earlier draft.

From the bottom of my heart, I thank Edward J. Taaffe and Howard L. Gauthier for their excellent instruction during my years at Ohio State University, and for helping me in every aspect of my career.

OnWord Press...

Dan Raker, President and Publisher
David Talbott, Director of Acquisitions
Rena Rully, Senior Manager of Editorial and Production
Barbara Kohl, Senior Editor and Book Editorial Division Manager
Carol Leyba, Senior Production Manager
Daril Bentley, Senior Editor
Lisa Levin, Senior Editor
Cynthia Welch, Production Editor
Michelle Mann, Production Editor
Kristie Reilly, Assistant Editor
Lynne Egensteiner, Cover designer, Illustrator

Contents

Chapter 2: Spatial Data   43

Chapter 3: Quantification of Spatial Analysis   81

Chapter 4: Single Layer Operations   119

Chapter 9: **Surface Analysis 309**

Chapter 10: **Grid Analysis**   **353**

Chapter 11: Decision Making in Spatial Analysis   387

Introduction

Spatial analytical techniques are dedicated to the analysis of the spatial order and associations of a phenomenon or variable. Spatial order delineates how geographic entities related to the phenomenon in question are organized in space, while spatial association describes the geographical relationships among phenomena. Therefore, a prerequisite of spatial analysis is that the study phenomenon must be mappable.

Maps provide an effective, two-dimensional representation of spatial distribution because the human eye is extremely adept at detecting spatial patterns. Consequently, the use of maps for detecting spatial relationships between map features is highly effective. As long as a need to divide and allocate space has existed, maps have been used as the primary instrument for the detection or imposition of spatial order. The fact that maps—analog representations of spatial order—were developed independently by numerous cultures throughout the world is proof of their value and significance.

Maps alone, however, are often not sufficient for analysis of spatial order and spatial association. Due to the complexity of spatial relationships among geographic entities, certain relationships (patterns) may be obscured or hidden in a generalized map. In addition, interpretation of a spatial relationship pattern by visual examination of a map is usually subjective, that is, perception of any map pattern may vary from person to person.

Recent advances in geographic information systems (GISs) have created a new arena for precision spatial analysis. A GIS combines the technologies of database management systems (DBMS) with automated cartography, and allows for map features to be geographically referenced with logically connected spatial data and map features. Consequently, entities related to the study phenomenon can be manipulated for thorough investigation of spatial patterns and relationships.

The most important contribution that new GIS technology has brought to spatial analysis is the establishment of a link between map-based analysis of spatial patterns and well-developed, rigorous quantitative analytical methods. With appropriate measurements of map features, interpretation of spatial patterns is no longer subjective. Because spatial patterns can be objectively assessed, hypotheses can be formulated and verified. In brief, cumbersome and time-consuming analysis of complicated spatial relationships has become increasingly accessible due to ongoing improvements in GISs.

If you have comments, questions, or recommendations about the book, please contact the author at OnWord press, 2530 Camino Entrada, Santa Fe, NM 87505-4835 USA, or email the publisher at *readers@bmp.com.*

Book Structure

The purpose of this book is to introduce the quantitative methods associated with GIS-based spatial analysis. Chapters 1 through 3 present basic concepts specific to GIS and spatial analysis. Chapters 4 through 10 focus on specific techniques and procedures. Chapter 11 addresses decision making guidelines to select a quantitative versus qualitative approach and the most appropriate methods for a given spatial analysis.

The glossary contains definitions of GIS, spatial analysis, and statistics terminology. Next, the references section lists numerous sources on methods and applications by chapter. Finally, the book ends with a detailed topical index.

Introduction to Spatial Analysis

Definition of GIS

A geographic information system (GIS) is an organized collection of computer hardware, software, geographic data, and personnel designed to efficiently capture, store, update, manipulate, analyze, and display all forms of geographically referenced information (Environmental Systems Research Institute, 1994). The primary purpose of a GIS is to process spatial, or geographically referenced, information. A subsidiary definition describes GIS as a computer system that stores and links geographically referenced data with graphic map features to allow a wide range of information processing, display operations, map production, analysis, and modeling (Antenucci et al., 1991). The conceptual framework of GIS, based on these definitions, is depicted in the following illustration.

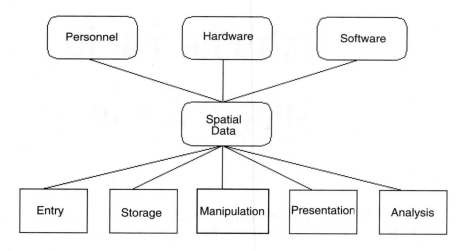

Conceptual framework of geographic information systems.

GIS Components

According to the conceptual framework described above, a GIS consists of four interrelated components: spatial data, personnel, hardware, and software. GISs are essentially designed for processing spatial data, the central component of any GIS. Any data considered spatial must be mappable, that is, every piece of information is attached to a specific object on a map, and the object's location on the map is geographically referenced.

Spatial Data

Spatial data come from a variety of sources. The most commonly used spatial data are obtained from sources such as the U.S. census, land use surveys, satellite imagery, aerial photographs, and paper maps. Recent technological advances in remote sensing and computer science provide access to tremendous amounts of spatial information on a daily basis. However, spatial data often contain variations in geographical coverage, data format, map scale, and data measurement. Integration of geographically referenced data from various sources is a fundamental characteristic of GIS software.

Traditionally, spatial data are presented in the form of hardcopy maps with pieces of information organized in separate document files. The traditional system of data management is not suitable for handling spatial data in the modern information age. First, hardcopy maps and document files require large amounts of storage space and are inefficient for organizing the huge amounts of spatial data that become available on a daily basis. Second, in order for data entry and retrieval to be effective, an index system is necessary to logically connect the geographical information on a map to its associated data in document files. Such an index system alone is difficult to maintain in hardcopy form; witness the multitudinous drawers storing index cards in the catalog room of public libraries.

The entry and retrieval efficiency for spatial data based on the traditional index system is no longer acceptable by today's standards. Consequently, most spatial data are now stored and organized in softcopy form (digital maps). For instance, the enormous quantity of spatial data transmitted from satellites is organized in a digital form for efficient storage and effective processing. The digital data stored on a single CD would require a large file cabinet in a map room if the spatial data were organized in hardcopy form. Next, digital data allow for much more effective data processing by computers. Indeed, GISs are developed specifically for utilizing state-of-the-art computing technologies to process spatial data in digital form. Spatial data properties and measurements, as well as the organization of spatial data, are the topics of Chapters 2 and 3.

Personnel

The effective use of a GIS typically involves complex procedures, and thus requires training and experience in several related fields. In effect, well-trained personnel are an indispensable component of GIS.

The growing number of certificate programs offered by GIS software vendors and educational institutions provides evidence of the recognized need for highly trained personnel.

Certification requirements vary from program to program. Principal qualifications of GIS personnel are listed below.

❐ Knowledge base of geography, cartography, and the computer and information sciences.

Geographic training provides the capability to evaluate spatial patterns and spatial processes, and to identify spatial relationships. Cartography offers knowledge of cartographic design and map making (e.g., map projections, coordinate systems, symbolization, and typography). Computer and information sciences provide general background in computer hardware, and skills for customizing daily operations through the development of suitable interface programs.

❐ Experience in the use of GIS software.

Major GIS software vendors offer intensive training courses focused on general GIS operations, macro language programming, and database management, among others.

❐ Thorough knowledge of the data.

Knowledge of the data encompasses source, precision and accuracy, original map scale and measurements of every data set, and of the organizational structure of GIS databases.

❐   Ability to conduct spatial analyses.

This capability requires training in spatial statistics and quantitative geographical methods. Such training permits the specialist to be able to select the best available analytical methods in an application, and to appropriately interpret the results.

The first three qualification areas for GIS specialists are covered in training offerings available through GIS software vendors and certificate programs of educational institutions. Training in the execution of spatial analysis is not as readily available. This book is designed for people who have already obtained basic training in GIS operations and need to use a GIS to conduct spatial analysis. Quantitative geographical methods appropriate to different types of spatial analyses are discussed throughout the book.

Hardware

Every GIS installation is comprised of the following hardware elements: the central processing unit (CPU), and devices for input, storage, and output.

CPU

The operating system, memory, and processing speed are the most important elements of a CPU. At present, most full-scale GISs operate on UNIX-based operating systems, while PC-based GISs tend to be of limited functionality. The mainframe systems that dominated the early stage of GIS history are being phased out, and their previous role is being assumed by UNIX-based workstations. UNIX systems may continue in the short run to serve as the main GIS units with full-scope functionality, and as servers for large databases. However, with the adoption of Windows NT, PCs can directly compete with UNIX systems.

A typical, effective system configuration consists of a small number of UNIX-based workstations installed with complete GIS software, such as ARC/INFO (Environmental Systems Research Institute-ESRI), for maintenance of the database and major GIS operations. The system also includes a larger number of viewing tools, such as ESRI's ArcView, for frequent data retrieval and basic spatial analyses. Current trends indicate that while the prices of UNIX-based systems continue to decline even as features continue to be upgraded, the performance of PC-based systems continues on an upward swing. Consequently, the price, memory, and speed gaps between workstations and PCs are narrowing. In the long run, it is likely that UNIX-based systems and PCs will converge.

Input, Storage, and Output Devices

Commonly used devices for data entry include digitizing tablets and scanners for converting analog data (in the form of hardcopy) into digital data. Tape drives and CD-ROM drives are required for retrieving data from existing sources. Storage devices vary greatly. The most frequently used storage media include hard disks, tape cartridges, and erasable optical disks. Output devices include printers for tabular data, summary reports, and results of analyses, and plotters for generating hardcopy maps and diagrams.

Competition among manufacturers continues to push hardware improvement and price reduction (especially for memory and CPUs). This trend encourages further development of GIS technology because enhanced hardware performance speeds up previously tedious and time-consuming operational procedures. Lower prices permit more users to afford computer equipment, thereby giving GIS developers incentive to continue research and development efforts. The recent substantial growth in networked environments suggests that the relative cost of hardware is falling because high-end equipment is shared among users. These trends indicate that the number of GIS users will continue to grow, as will the need for spatial analysis.

Software

The processing of spatial data requires two interrelated software components: automated cartography and database management. Modern GIS technology evolved from recent advances in the development of these two components.

Automated Cartography

Cartography is the science, art, and technology of making maps. Therefore, automated cartography is mapmaking with the use of computer equipment. A map is a graphic representation of spatial relationships and forms (Robinson et al, 1984), and every map is a scaled-down model of the reality. The generation of such a model requires cartographic abstraction, that is, unmapped data of the reality must be transformed into map form. Four procedures are essential in cartographic abstraction: selection, classification, simplification, and symbolization (Dent, 1990).

Computers assist cartography in several aspects. First, because map data are in digital form, errors on a map can be easily corrected and a revised map can be produced with much less effort than without computers. Although digitizing may introduce data errors and reduce accuracy, such errors can be effectively corrected once detected. In this regard, the precision of map information can be significantly improved. Furthermore, updating map data becomes a much easier task.

Second, the process of cartographic abstraction can be carried out at a higher speed and lower cost. The selection, classification, and simplification of map features can be executed scientifically. Procedures of cartographic abstraction and generalization can be programmed and built into a package. Consistent results can be obtained by different cartographers or by the same cartographer at different times.

Third, cartographic design can be greatly improved through trial and error. The size, shape, or position of text or symbols on a map can be quickly changed and effectively corrected.

Database Management

The second software component of GIS is an efficient database management system (DBMS). GISs must have the capability of manipulating varying items of geographically referenced data. Because database volume is typically enormous, the DBMS must be able to handle large amounts of data efficiently. In addition to the large amounts of data associated with places, geographical information describing each place must also be recorded.

The most important element of a GIS is the logical link between an automated cartography component and the DBMS. Descriptive data associated with each individual place must be logically connected with the spatial information of that place. The logical connection between attribute data and map features is essential for GIS operations.

First, editing of the attribute data must be reflected in the map features. Thus, if the population of one county is revised, the revision of the map showing county population must be automated, and no additional effort should be required.

Without a GIS, the cartographic data are separated from attribute data, and thus revision of the attribute is not automatically incorporated in the cartographic data. Consequently, additional efforts are needed in order to revise map features according to changes in attribute data.

Second, changes to map features must also be automatically incorporated into the attribute data. For instance, if the boundary between two counties is revised to correct previous errors, the effects of the revised boundary on the attribute data must be entered into the attribute table automatically, without the need for additional efforts. In this simple example, the area of each county affected by the revision is computed automatically.

GIS Software Functionality

At the early stages of development, GISs were envisioned to provide functions such as acquiring, storing, manipulating, and displaying geographic data for decision making (Calkins and Tomlinson, 1977). Accordingly, GIS basic functionality can be summarized as follows:

❐   Entry/updating

Because GISs are designed for spatial data processing, the most fundamental feature of a GIS is data entry and updating. Data entry and updating are inseparable operations; any system that allows data entry into the database must also allow for database updating. There are numerous desktop mapping systems available on the market. Any system that does not provide a data entry/updating function cannot be considered a GIS because this function is a prerequisite to all other types of GIS operations.

Data entry and updating must also permit the use of source data in either digital or analog form. The spatial information obtained from a paper map or aerial photograph should be translated into a digital format, and digital data of different sources and formats should be converted into a format compatible with the database.

❑   Data conversion

Data conversion is a function closely related to data entry and updating. Because most commercial GISs use built-in proprietary data formats and data models, competing vendors are reluctant to release their internal data formats and data models in order to protect trade secrets. On the other hand, GIS users must avoid unnecessary, redundant efforts at digitizing already existing data. The same digital data set typically can be found in numerous GIS data formats.

Furthermore, government spatial data sets may not be organized in a single GIS format. For instance, U.S. Census data and corresponding spatial information in the Census Bureau's TIGER/Line file, and the U.S. Geological Survey's digital elevation models (DEMs) and digital line graphs (DLGs) are available in different formats. These data are useful for a variety of applications. Thus, major GIS software manufacturers provide utilities for the conversion from one data format to another.

❑   Storage/organization

The importance of storage and organization functions in a GIS is due to the variability of spatial data. The data that GIS deal with are not only large in volume, but also of great variability. As mentioned previously, spatial data vary in geographic coverage, data format, variable measurements, and map scale. Such a variety of data must be efficiently stored and organized. Two important properties are required for a GIS to be efficient in storage and organization. First, spatial information for different locations or geographic coverages must be organized to allow for accurate and effective retrieval of information for any desired location or geographic coverage. Second, spatial data for the same locale that vary in scale, format, or measurement must be logically connected for effective processing.

❐ Manipulation

Because many GIS operations require spatial data to be selected by certain criteria, classified in different ways, combined to form new variables, and altered for model building, GISs must provide the capability of manipulating spatial information. This function must allow users to classify and identify map features through either query statements or interactive graphic inputs. Different attributes can be combined in any specified manner and new variables can be defined, computed, and altered.

❐ Presentation/display

In the same way that every computer application requires output, GISs must also provide an efficient tool for the presentation of spatial data. Spatial data, either retrieved directly from the database in its original form or processed through manipulation procedures, can be presented in text format, tabular format, or map format. Summary statistics and analysis results are typically stored in text files for printer output or exchange between different softwares or operating systems. Selected entries and attributes of the database can be displayed in tables or fixed format files. Maps are generated for display on the monitor or saved in a plot file for plotting. In this regard, a GIS must supply necessary output drivers for efficient display and presentation of data.

❏ Spatial analysis

In the early stages of GIS development, only the five functions mentioned above were envisioned by GIS developers. A sixth required function became apparent as soon as the technology advanced to a degree where GIS became useful for empirical applications. The earlier definition of GIS effectively overlooked the direct use of GIS in spatial analysis. According to ESRI's definition, in addition to others mentioned above, a complete GIS must provide spatial analysis functions. Indeed, it is the spatial analysis capability that differentiates GIS from desktop mapping software. In general, a complete mapping package may provide all GIS functions with the exception of spatial analysis. Hypothetical distributions can be generated with specified parameters, while descriptive, explanatory, and predictive models can be derived.

Using GIS for Spatial Analysis

Spatial analysis in GIS involves three types of operations: attribute queries (also known as aspatial queries), spatial queries, and generation of new data sets from the original database. The scope of spatial analysis ranges from a simple query about the spatial phenomenon to complicated combinations of attribute queries, spatial queries, and alterations of original data.

In GIS applications, attribute and spatial queries are common. *Attribute queries* require the processing of attribute data exclusive of spatial information. For instance, from a database of a city parcel map where every parcel is listed with a land use code, a simple attribute query may require the identification of all parcels for a specific land use type. Such a query can be handled through the table without referencing the parcel map. Because no spatial information is required to answer this question, the query is considered an attribute query. In this example, the entries in the attribute table that have a land use code identical to the specified type are identified. Further information can then be generated, such as the number of parcels of this land use type, or the total area of this land use type in the city.

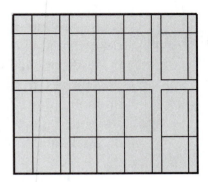

A sample parcel map.

The previous illustration shows a sample parcel map in which each parcel is a record in a database. Attributes in the database include parcel number, parcel size, value, land use, and so on. The following table presents part of the database.

Attribute table of the sample parcel map

Parcel number	Size	Value	Land use
102	7,500	200,000	Commercial
103	7,500	160,000	Residential
104	9,000	250,000	Commercial
105	9,000	240,000	Industrial
106	7,500	260,000	Commercial
107	6,600	125,000	Residential

A typical query involves identifying commercial land use parcels in order to compute the average value of this land use type. The selection is based only on an attribute item; therefore, no spatial information is required.

Spatial queries require the processing of spatial information. For instance, a question may be raised about parcels within one mile of a major freeway. The answer to such a query requires spatial information about the location of the freeway and each parcel. In this case, the answer can be obtained either from a hardcopy map or by using a GIS with the required geographic information.

The next illustration shows a spatial query example. Assume that when a request is submitted for rezoning, all owners whose land is within a certain distance of the parcel that may be rezoned must be notified for public hearing. A spatial query is required to identify all parcels within the specified distance. Through a spatial query, all the parcels within the specified distance can be identified. This process cannot be accomplished without spatial information. In other words, the attribute table of the database alone does not provide sufficient information for solving problems that involve location.

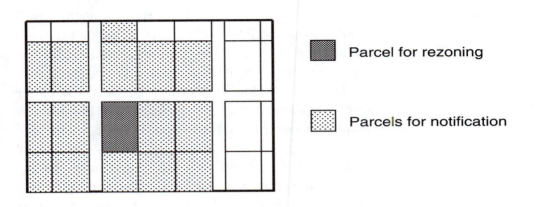

Landowners within a specified distance from the parcel to be rezoned identified through a spatial query.

In most cases, attribute queries and spatial queries can be handled with a GIS when the question requires no data alterations. A spatial analysis, however, may require generation of new data from the original set. For instance, assume that you wish to find the relationship between noise level and proximity to freeways (e.g., within one mile of the freeway) for all parcels of a specific land use code (e.g., residential land use). To answer this question, you must combine (i.e., overlay) the parcel map, the freeway map, and the areas within one mile of the freeway derived from the freeway map. This process may require the delineation of new geographic entities and the generation of a new data table showing combinations of several factors.

The following illustration shows an example in which residential land parcels within a certain distance from freeways are identified for a spatial analysis of noise pollution. Assume that residential parcels beyond the specified distance are classified into a different category. Comparisons can be made between these two types of residential parcels with respect to noise level. A discernible difference between them will indicate that proximity to freeways is a significant factor contributing to noise pollution in residential neighborhoods.

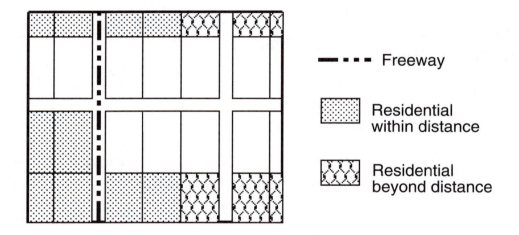

Parcels of residential land use within a specified distance from freeways are identified for noise pollution analysis.

While basic spatial analyses involve some attribute queries and spatial queries, complicated analyses typically require a series of GIS operations including multiple attribute and spatial queries, alteration of original data, and generation of new data sets. The methods for structuring and organizing such operations are major concerns in spatial analysis. An effective spatial analysis is one in which the best available methods are appropriately employed for different types of attribute queries, spatial queries, and data alteration. The design of the analysis depends on the purpose of the study. In general, every spatial analysis must start with an explicitly addressed research question which requires the processing of spatial information. All effort expended in the analysis is structured toward answering the question.

Significance of Spatial Analysis

Spatial analysis is one of the most important uses of GIS. Since the advent of GIS in the 1980s, many government agencies have invested heavily in GIS installations, including the purchase of hardware and software and the construction of mammoth databases. Two fundamental functions of GIS have been widely realized: generation of maps and generation of tabular reports. Indeed, GISs provide a very effective tool for generating maps and statistical reports from a database. However, GIS functionality far exceeds the purposes of mapping and report compilation. In addition to the basic functions related to automated cartography and database management systems, the most important uses of GISs are spatial analysis capabilities. As spatial information is organized in a GIS, it should be able to answer complex questions regarding space.

The huge investment in hardware, software, and database construction among major government agencies has resulted in numerous maps and corresponding report data. However, questions are raised as to why so much investment should be put into GIS if the same job can be accomplished with less costly mapping and statistical packages. GIS specialists have gradually realized that the most important use of GIS is spatial analysis, the specific GIS functionality that mapping and statistical packages cannot provide.

Making maps alone does not justify the high cost of building a GIS. The same maps may be produced using a simpler cartographic package. Likewise, if the purpose is to generate tabular output, then a simpler database management system or a statistical package may be a more efficient solution. It is spatial analysis that requires the logical connections between attribute data and map features, and the operational procedures built on the spatial relationships among map features. These capabilities make GIS a much more powerful and cost-effective tool than either automated cartographic packages, statistical packages, or database management systems. Indeed, functions required for performing spatial analysis that are not available in either cartographic packages or database management systems, are commonly implemented in GIS.

GIS Usage in Spatial Analysis

GIS operational procedures and analytical tasks that are particularly useful for spatial analysis include single layer operations, multiple layer operations, spatial modeling, point pattern analysis, network analysis, surface analysis, and grid analysis.

● *Single layer operations* are GIS procedures which correspond to attribute queries, spatial queries, and alterations of data that operate on a single data layer.

The illustration below shows a single layer operation example where buffer zones are generated from streets. Buffer zones are spatial expansions of point, line, or polygon features. In this example, streets are represented by solid lines. A GIS is used to build a buffer of a given distance extending outward from every street segment within a specific area.

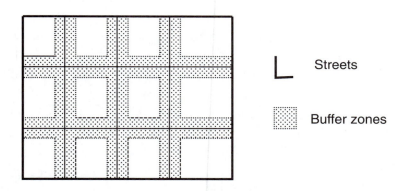

└ Streets

⬚ Buffer zones

Buffer zones extended from streets.

● *Multiple-layer operations* are useful for manipulation of spatial data on multiple data layers.

The next illustration shows two data layers on the left side, one representing a map of soil polygons and the other, a land use map. Soils are classified into three categories labeled 101, 102, and 103. Land use types include forest (F), agricultural (A), and urban (U). The overlay of these layers produces the new map on the right in which polygons of different combinations of soil and land use are delineated. Each polygon in the combined map is different from any surrounding polygon in terms of soil or vegetation, or both. For instance, the original soil polygon 101 is now divided into two smaller polygons because of different land use types.

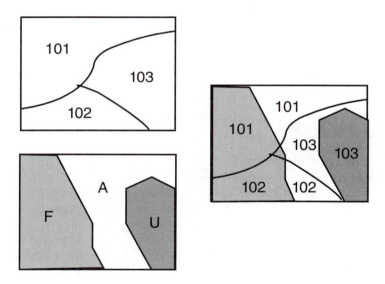

The overlay of two data layers creates a map of combined polygons.

● *Spatial modeling* involves the construction of explanatory and predictive models for statistical testing.

The illustration below shows an example of air pollution spatial modeling. Emissions of a specific particulate are measured at monitoring stations represented as point locations on the bottom layer. The distribution of air pollution is believed to be related to soils (silt content and other soil characteristics), agricultural operations, roads, and topography. With a database containing all required data elements, a spatial model can be constructed to explain the distribution of air pollution based on these related variables.

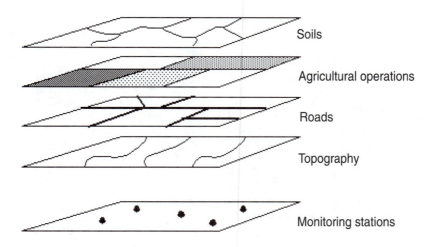

A spatial model of air pollution based on the distributions of several variables.

- *Point pattern analysis* deals with the examination and evaluation of spatial patterns and the processes of point features.

A typical biological survey map is shown below in which each point feature denotes the observation of an endangered species such as big horn sheep in southern California. The objective of illustrating point features is to determine the most favorable environmental conditions for this species. Consequently, the spatial distribution of species can be examined in a point pattern analysis. If the distribution illustrates a random pattern, it may be difficult to identify significant factors that influence species distribution. However, if observed locations show a systematic pattern such as the clusters in this diagram, it is possible to analyze the animals' behavior in terms of environmental characteristics. In general, point pattern analysis is the first step in studying the spatial distribution of point features.

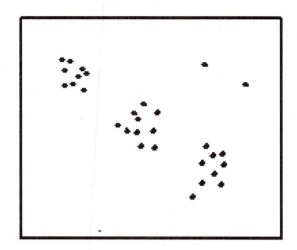

Distribution of an endangered species examined in a point pattern analysis.

- *Network analysis*, designed specifically for line features organized in connected networks, typically applies to transportation problems and location analysis.

The next illustration shows a common application of GIS-based network analysis. Routing is a major concern for the transportation industry. For instance, trucking companies must determine the most cost-effective way of connecting stops for pick-up or delivery. In this example, a route is to be delineated for a truck to pick up packages at five locations. A routing application can be developed to identify the most efficient route for any set of pick-up locations. The highlighted line represents the most cost-effective way of linking the five locations.

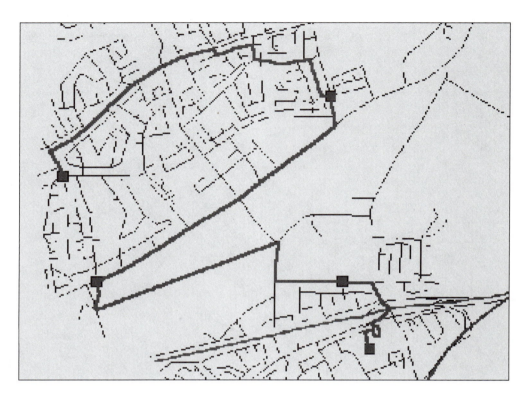

The most cost-effective route links five point locations on the street map.

- *Surface analysis* deals with the spatial distribution of surface information in terms of a three-dimensional structure.

The distribution of any spatial phenomenon can be displayed in a three-dimensional perspective diagram for visual examination. A surface may represent the distribution of a variety of phenomena such as population, crime, market potential, and topography, among many others. The perspective diagram of Cajon, California, appearing below was generated from the U.S. Geological Survey's digital elevation model (DEM) through a series of GIS-based operations in surface analysis.

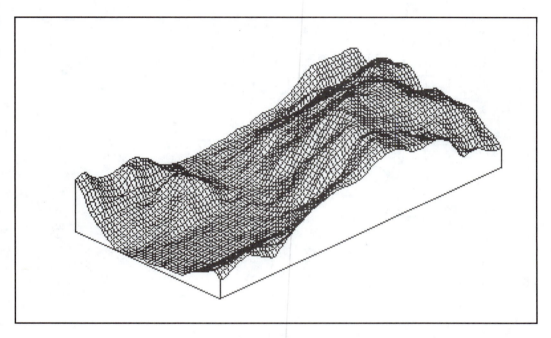

Perspective diagram of Cajon, California derived from a surface analysis.

● *Grid analysis* involves the processing of spatial data in a special, regularly spaced form.

The following illustration shows a grid-based model of fire progression. The darkest cells in the grid represent the area where a fire is currently underway. A fire probability model which incorporates fire behavior in response to environmental conditions such as wind and topography delineates areas that are most likely to burn in the next two stages. These areas are represented by lighter shaded cells. Fire probability models are especially useful to fire fighting agencies for developing quick-response, effective suppression strategies.

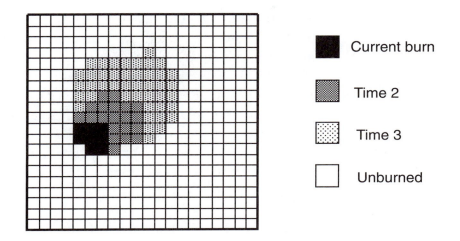

A fire behavior model delineates areas of fire progression based on a grid analysis.

In most cases, GISs provide the most effective tool for performing the above tasks. Each of the tasks is a topic of detailed discussion in subsequent chapters.

Spatial Analysis Applications

GIS applications have been employed in many areas. In environmental research, GISs are used to map the distribution of forest cover, land use, natural habitats, natural resources, environmental hazards, and much more. In wilderness area fire management, different layers of spatial information such as vegetation, soil, hydrology, and human activities are organized in a GIS, and critical areas of fire danger are delineated.

Probabilities of fire occurrence are mapped and statistically tested. Possible spatial strategies are then evaluated and compared for more effective wildfire management.

Government agencies have recently adopted a variety of GIS-based spatial analytical techniques to deal with public health issues. For instance, reported cases of AIDS are entered into GIS databases for both monitoring and modeling purposes. Geographic distributions of disease victims are analyzed periodically in order to detect patterns in disease outbreaks. Modeling efforts often emphasize the identification of possible relationships between disease distribution and socioeconomic characteristics.

In the area of air quality control, the South Coast Air Quality Management District (SCAQMD) in southern California has recently launched a GIS project for modeling the distribution of particulate matter under 10 mm (PM10). Selected emission sources were identified for analysis, including construction activities, unpaved roads, and land use, among others. Efforts have been undertaken to build a comprehensive database and develop macro language programs using ESRI's ARC/INFO GIS. Eventually, the observed data of PM10 emissions collected at monitoring stations and contributing factors from identified sources will be incorporated in a spatial model. In the future, not only will critical information about emissions be released to the public on a real-time basis, but the most effective management measures of pollution reduction can also be implemented to minimize PM10 emissions.

A sample of municipal GIS applications includes land use mapping, rezoning modeling, mapping spatial distribution of the population, school bus routing, urban transportation planning, and demographic analysis. For law enforcement agencies, pin maps of crime incidence and crime reports, previously handled manually, now can be much more effectively manipulated using GIS application tools. Crime analysis using state-of-the-art technologies is especially helpful for law enforcement agencies in crime prevention and control of gang activity.

Private sector GIS applications are as multifaceted as private sector activities. For instance, the release of census data, along with the nationwide street data in the U.S. Census Bureau's TIGER/Line file, has enabled marketing firms to evaluate market potentials based on actual distributions of demographic and socioeconomic characteristics. For insurance companies, geographically differentiated rates of flood and auto insurance can be more accurately calculated so that more competitive rating schemes can be established. Available demographic and infrastructure databases, integrated with built-in functions of data presentation and spatial analysis, provide utility companies with very powerful tools for system maintenance and strategic planning.

Appearing below are a few empirical examples of spatial analysis applications.

Analysis of Landscape Connectivity

The purpose of a landscape connectivity analysis project conducted at the University of California, Riverside, and sponsored by the California Department of Forestry and Fire Protection (CDFFP) is to develop GIS methods for evaluating an ecosystem's degree of connectivity. Several timber harvest plans were submitted by private companies to CDFFP for evaluating possible impacts of each proposal on the ecosystem in eastern California's central Sierra Nevada. Given the current distribution of forest cover, CDFFP wants to develop a method for deriving measures of landscape connectivity to determine the long-term effects of timber harvest plans on the landscape.

An ARC/INFO macro language program was developed to accomplish the following tasks:

● Classify the landscape into categories pertaining to landscape connectivity.

● Set spatial filters to adjust geographic units located around the edge of a cluster of late successional forests.

● Provide spatial constraints to reclassify isolated clusters under a minimum size.

● Compute indices related to landscape connectivity.

The next diagram illustrates a simplified task of landscape connectivity analysis. A grid is constructed to divide the forested area into uniform units (grid cells). The units are classified into four habitat categories for an endangered species. The darkest cells represent core areas of natural habitat suitable for nesting; dark cells denote general activity space; light cells represent corridors where the animals can move relatively freely; and unshaded cells are considered barriers to the species such as freeways and developed areas. The evaluation of landscape connectivity is concerned with the extent to which core habitat cells are connected through activity space and corridors. For land management agencies, it is important to maintain the minimally required level of landscape connectivity in order to ensure the survival of the species in question.

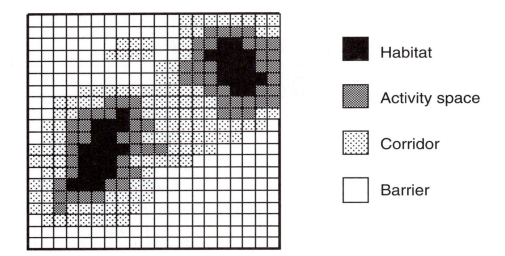

Landscape connectivity is evaluated by the degree to which core habitat cells are connected through activity space and corridors.

The computer program was developed and additional research conducted to develop a quantitative method using linear programming tools for the optimization of landscape management. The same method can be used by CDFFP to determine the most cost-effective spatial strategies in managing forest and wildland fires. In general, the method is applicable to other projects requiring the evaluation of long-term environmental management plans.

Modeling Wildfire Distribution

A series of related projects sponsored by the U.S. Forest Service is aimed at developing GIS operational procedures and research methods for effective wildfire management. A comprehensive database was constructed to include multiple layers of environmental and human-related variables for the San Jacinto Ranger District of the San Bernardino National Forest, California. Variables derived from the database encompass fire history, vegetation, topography, soils, human structures, hydrology, and climatic conditions.

The database was then used to test a model of fire occurrence probability. The model estimates the probability of a major fire by geographic unit, and results were used to construct a map showing the distribution of wildfire occurrence probability. Important applications of the model include the delineation of critical fire danger zones, evaluation of alternative spatial strategies in preventive treatments, and identification of the most cost-effective management plans.

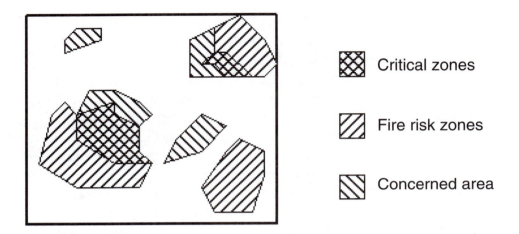

Critical zones are defined as high fire risk areas.

The previous illustration shows partial results of wildfire modeling. Spatial models were built to delineate areas of high fire occurrence probability, represented as polygons of high fire risk. In addition, areas containing human properties, habitat of endangered species, and important natural resources are identified through GIS operations. The overlay of these two layers, fire risk and concerned areas, produces a map delineating critical zones of fire danger. Critical zones are areas that either require special treatment during a fire or preventive measures at the planning stage. The fire probability model provides fire management agencies a practical, useful tool for several management tasks.

Analysis of Air Transportation Accessibility

This study examined the change in accessibility to major air transportation services in the United States since the deregulation of the airline industry. For each major U.S. city, the average cost of traveling to all other cities is estimated based on actual data of inter-city airline traffic. The evaluation of average travel cost takes into consideration the network structure of hub-and-spoke operations employed by major domestic carriers. Average cost comparisons were made for 1970 and 1989. The differences were mapped to show spatial patterns and relationships between these changes, and changes in population growth and income distribution. Changes in spatial pattern were statistically evaluated using multiple regression analysis.

Changes in air transportation accessibility in the United States between 1970 and 1980.

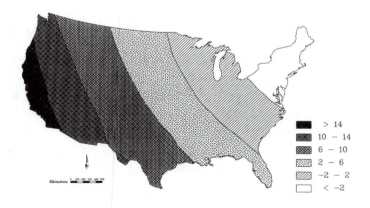

> 14
10 – 14
6 – 10
2 – 6
–2 – 2
< –2

The previous illustration shows a simplified version of the changes in accessibility to domestic air transportation in the United States between 1970 and

1980. Major cities were evaluated in terms of air transportation accessibility for both years. Using the spatial interpolation functions in the ARC/INFO GIS, the changes in air transportation accessibility were translated into zones of similar pattern. The country was divided into five zones. The west coast enjoyed the largest increase in air transportation accessibility while the northeastern states suffered the largest relative decline. The spatial pattern suggests that during the study decade, there was a clear trend in airline traffic toward the west coast. The pattern became more complicated in the 1980s.

School Bus Routing

A GUI developed for a school district in southern California specializes in school bus routing. The system consists of the following interrelated modules:

- *Manual routing.* Allows the user to pick and sequence street segments and define a route, or to pick a sequence of bus stops and allow the shortest path route to be automatically delineated.

- *Automatic routing.* Reads a table of geocoded passenger addresses and identifies the best routing schemes for the school.

- *Passenger plotting.* Reads the passenger list, finds the location of each passenger on the street map, and plots the map showing both streets and passenger locations.

● *Walking distance.* Allows the user to identify a school and set the maximum distance within which students will not be provided with free bussing services. The module then finds all street segments within the specified walking distance.

● *Bus stop optimization.* Reads all passenger addresses, and from the street map, finds the optimal locations for bus stops.

● The special education module is designed specifically for students in special education programs. Busses equipped with special facilities such as wheelchair slots are allocated to the routes.

The following illustration shows the distribution of students generated from the geocoding function based on the master file of student addresses. The student file is imported into the system for geocoding. Each record contains detailed information about a particular student, including his/her address. An automated geocoding procedure reads the file and identifies the location of each address, based on the U.S. Census Bureau's TIGER/Line street file.

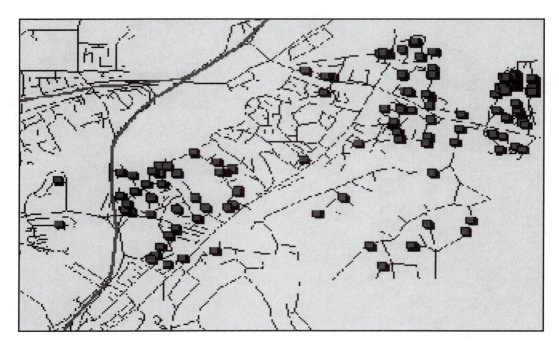

Student locations are generated through address geocoding.

Street locations are identified for bus stops according to student address distribution as shown in the next illustration. All students are assigned to the nearest respective bus stops.

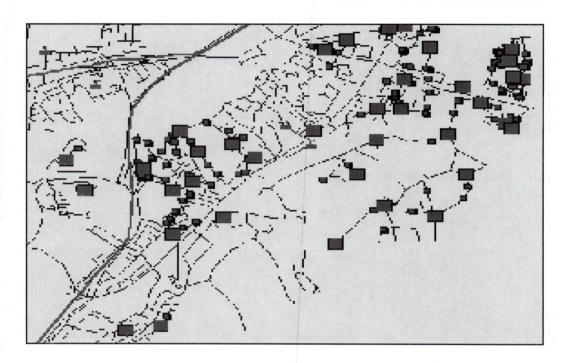

Bus stop locations (large squares) are identified according to student address distribution.

At each stop, the total number of students that require bussing services is computed. Depending on the number and seating capacity of buses, multiple routes are delineated to connect the bus stops in the most cost-effective way. The diagram below illustrates a sample set of bus routes.

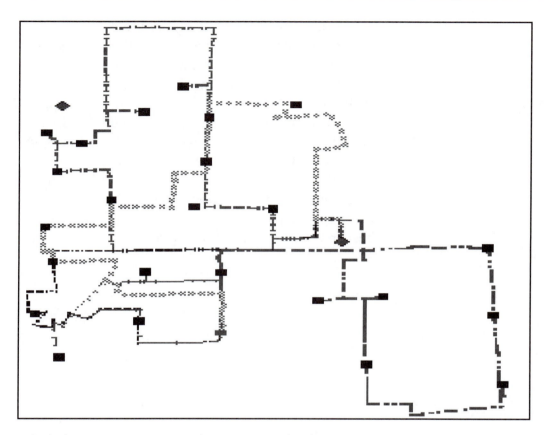

Multiple bus routes connecting bus stops to schools.

This case is a typical application of spatial analysis in which the procedures for finding a specific solution are programmed for a minimal user interface.

Summary

This chapter provides an overview of GIS-based spatial analysis. GISs are useful for manipulation, presentation, and analysis of geographically referenced data. The full potential of spatial analysis based on available GIS functionality remains to be discovered;

many other applications are possible. GIS functionality and major components are outlined to provide the reader with a broad, yet fundamental, understanding of the requirements and usefulness of GIS technology. Empirical cases of GIS-based spatial analysis, including landscape connectivity analysis, wildland fire modeling, air transportation accessibility, and school bus routing are briefly described to provide typical examples of GIS applications. In brief, spatial analysis capabilities make GISs powerful tools in the activities of government agencies, research institutions, and private corporations.

Spatial Data

A spatial analysis is designed to answer questions pertaining to the spatial order and/or spatial association of a phenomenon. Thus, a spatial analysis always has explicitly expressed objectives. The type of spatial features under consideration and the nature of the problem collectively determine both the data elements and available methods required for the analysis. In general, spatial analyses can be classified into five categories based on data requirements: (1) point pattern analysis for point features; (2) network analysis for line features; (3) spatial modeling for polygon features; (4) surface analysis for volumetric data; and (5) grid analysis for regularly spaced data. Each type of spatial analysis is treated in subsequent chapters.

Spatial data are the central component of a GIS and the building blocks of a spatial analysis. Spatial analyses dealing with different types of problems require the use of spatial data with correspondingly different characteristics. The properties of spatial data and the organization of the various data elements in a GIS are the focus of this chapter. Considerations to be addressed include the following:

● Differential spatial data organization

● Definition and map representation of cartographic objects

● Construction of various data elements for specific data types

● Differences in data structure

The first section focuses on the two general types of data analysis, spatial and aspatial, that are associated with different data requirements. Because every spatial analysis deals with spatial forms and spatial relationships, the objects involved in the analysis must be representable on a map. How such objects are represented in a database is the focus of the second section. The third section of the chapter presents more detailed discussion of the required data elements. Two distinctive data models, *raster data structure* and *vector data structure*, determine both the characteristics of data elements and the operational procedures of a GIS. Finally, the *arc-node data model*, which is widely adopted by the manufacturers of major vector based GISs, is described.

Spatial and Aspatial Analyses

Quantitative data analysis is either spatial or aspatial. A spatial analysis is focused on the role of space and relies on explicitly specified spatial variables in the explanation or prediction of a phenomenon. For aspatial analyses, spatial components and spatial information are not required for analysis and explanation, respectively. The key difference between these two types of data analysis is the inclusion or exclusion of spatial factors. In short, all spatial analyses involve processing information about geography.

Every phenomenon is directly and indirectly related to variables. Some variables may have more significant geographic implications than others. The same phenomenon can be analyzed spatially or aspatially, depending on whether the subject of interest involves a geographic relationship.

For instance, census data are geographic in nature, but can be used in aspatial analyses such as examining the relationship between income and education without any reference to geography. The latter study is aspatial because no geographic information is required in processing. If the relationship between income and education is compared for different regions, however, the analysis becomes spatial because geographic units must be divided into regions.

Although spatial analyses always require variables with geographic implications, aspatial analyses may or may not involve such variables. Examples of variables that do not have direct geographic implications include weight, height, income, and education. Variables with clear geographic implications include location, accessibility to transportation, proximity to essential facilities, and so forth.

The following illustration shows an example data set employed in aspatial analyses. The rows in the table represent records of an object such as an individual, household, county, and so forth. Columns represent attributes or variables. In this example, the table contains the data of three attributes for five individuals. The attributes are age, education, and income, none of which has any significant geographic implications. Typical questions for the sample depicted below include the relationship between income and age, or the relationship between income and education. A typical aspatial attribute query would be the selection of all subjects 40 years of age or older.

		Attributes		
	ID	**Age**	**Education**	**Income**
Objects	1	35	12	49
	2	43	16	66
	3	26	16	32
	4	62	12	58
	5	47	20	62

Objects in spatial analyses must be mappable, and thus, geographically referenced. The objects on a map can be represented as point, line, or polygon features. The data set must contain information about the position of every feature typically represented by x and y coordinates. The following table shows an example of a data set suitable for spatial analysis. In this case, the five point locations denoted by longitude and latitude allow for each location to be geographically referenced for presentation on a map. Each location has an additional variable of spatial relevance (distance from the coast), and a variable for analysis (temperature).

Attributes				
ID	**Longitude**	**Latitude**	**Distance**	**Temperature**
1	121	48	23	24
2	132	37	44	32
3	78	68	32	21
3	78	68	32	21
4	69	24	15	30
5	113	42	36	26

Objects

Based on the above data set, a spatial analysis can be conducted to answer questions about the relationship between latitude and temperature, or the relationship between temperature and distance from the coast. The same data set could be used for both aspatial and spatial queries. For instance, a possible aspatial query would be the selection of all locations with temperatures under 30 degrees, whereas spatial queries would be the selection of all locations within 30 miles of the coast or higher than 30 degrees in latitude.

To demonstrate the differences between spatial analysis and aspatial analysis, consider the phenomenon of income distribution. A sociologist or economist may conduct a study of changes in U.S. household income distribution between 1985 and 1995. If the researcher is exclusively concerned with whether the gap between the wealthiest and the poorest increased or decreased over the ten-year period, no spatial components are involved and the analysis is aspatial. An aspatial analysis of income distribution could employ census data for all households in City X metropolitan area without reference to geography. City X is simply a population group, not a place. In this scenario, households could be divided into five quantile groups (from the highest fifth to the lowest fifth) for 1985 and 1995. The analysis may be focused on the change in the gap between the top and bottom quantiles from 1985 to 1995. Appearing below is a fictitious rendering of the aggregate results of such a study.

Household income by quantile, 1985

Quantile	Mean household income	Total income by quantile	% of total household income
Highest 20%	$65,000	$2.275 billion	32.5
20%	$53,000	$1.855 billion	26.5
20%	$39,000	$1.365 billion	19.5
20%	$24,000	$840 million	12.0
Bottom 20%	$19,000	$665 million	9.5

Total number of households = 175,000. Total household income = $7 billion.

Household income by quantile, 1995

Quantile	Mean household income	Total income by quantile	% of total household income
Highest 20%	$73,000	$2.92 billion	34.4
20%	$58,000	$2.32 billion	27.4
20%	$38,000	$1.52 billion	17.9
20%	$25,000	$1 billion	11.8
Bottom 20%	$18,000	$7.2 million	8.5

Total number of households = 200,000. Total household income = $9.48 billion.

Alternatively, a geographer may conduct a spatial analysis to examine the change in income distribution between 1985 and 1995 with income distribution defined geographically. In this case, household income data aggregated at a certain level must be mapped in order to show spatial distribution. Assume that the study in question is focused on the same City X as the previous example, and the level of aggregation is census block group. Examination of the census data now requires spatial components. A typical spatial analysis can be conducted in such a way that all census block groups are classified into five quantile categories according to mean household income in both years. Maps showing the distribution of the income categories for 1985 and 1995 can then be compared.

The next illustration shows a typical spatial analysis of income distribution in a fictitious area. The spatial pattern in the change of income distribution between 1985 and 1995 can be detected by comparing the two maps. In 1985, the highest income groups are clustered near the north central region of the district while lower income groups are dispersed outward from high income clusters. In 1995, part of the northern portion of the district gained locational advantage in income distribution whereas the southern portion suffered economic deterioration.

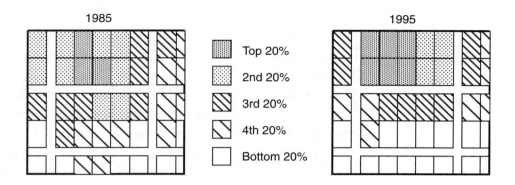

1985
1995

Top 20%

2nd 20%

3rd 20%

4th 20%

Bottom 20%

1985 and 1995 income distributionsin a sample district.

Spatial analysis differs from aspatial analysis in both the required data components and analytical tools. In the study of income distribution, for instance, the aspatial analysis does not require processing geographic variables; therefore, a database management system (DBMS) and a statistical package provide sufficient analytical capability for the aspatial analysis. In the spatial analysis, the processing of spatial information makes analytical tasks more complicated, and usually a fully functional GIS is required.

Representation of Spatial Objects

In a typical spatial analysis, the spatial objects linked to the study phenomenon must be represented on a map in order to construct their spatial relationships. The data for both locations and attributes are processed through a series of programs in a GIS. A prerequisite to efficient data handling is that spatial objects be defined in a consistent and unambiguous manner. In terms of digital representation, spatial objects are classified into point, line, and area (polygon) features. The following illustration shows selected cartographic objects defined by the National Committee for Digital Cartographic Data Standards (NCDCDS). A point feature is a zero-dimensional object that specifies geometric location. Therefore, point features only represent positions; they have no other meaningful measurement. Even though point symbols may vary in size, the area of such symbols is meaningless.

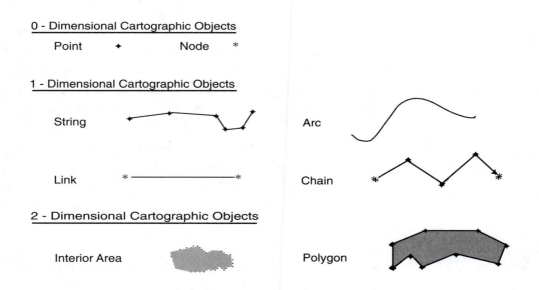

Selected cartographic objects defined by NCDCDS.

The four types of point features defined by the NCDCDS follow.

- *Entity point.* Used to identify the location of point features such as buildings, wells, and utility poles. As such, the accuracy of the location of entity points is important.

- *Label point.* Used for displaying text associated with map features. For this type of point, the position accuracy is solely a cartographic concern. In brief, the analyst must verify that the annotation is attached to the correct object and that map viewers will not be confused by the position of one label vis-a-vis another label.

- *Area point.* Denotes a position within an area feature, and is used to carry the attribute information associated with the area feature. For instance, a point representing the county seat may be assigned the statistics for the entire county.

- *Node.* Represents a point location with additional topological significance such as intersections or end points of line features.

Point features are one-dimensional in spatial analysis, although they represent two-dimensional objects on maps. For instance, one point may represent a well, while another represents a utility pole. Although the area that a well occupies is different from that of a pole, on a map covering an entire city the difference between the area of a well and a pole is negligible due to map scale. In a spatial analysis, the difference in area among different point features is considered meaningless.

Line features are one-dimensional and represent both position and direction. Length is a significant measurement of line features. Although linear features actually occupy two-dimensional space on maps, their width is not considered in cartographic renderings. For instance, roads and rivers are commonly represented as line features. This representation is meaningful only in terms of length, although their actual measurements are both length and width. Physical objects are conventionally represented by solid lines while intangible features of one-dimensional objects such as political boundaries are often represented by dashed or broken lines.

According to the NCDCDS, the following terms are specified for *line features*. (See previous illustration for display of selected line features.)

● *Line.* Generally defined as a one-dimensional object.

● *Line segment.* Direct line between two points.

● *String.* Sequence of line segments without nodes, node identifiers, or left and right identifiers. A string may intersect itself or other strings.

● *Arc.* Locus of points that forms a curve defined by a mathematical function.

● *Link.* One-dimensional object serving as a connection between two nodes. Links are also known as *edges*.

- *Directed link.* Link between two nodes with a specified direction.

- *Chain.* Directed sequence of non-intersecting line segments and/or arcs with nodes at each end. References to left and right area identifiers are optional.

Area features are two-dimensional and represent both position and area. NCDCDS definitions of area feature terms follow.

- *Area.* Bounded, continuous, two-dimensional object which may or may not include its boundary.

- *Interior area.* Area not including its boundary.

- *Polygon.* Area consisting of an interior area, one outer ring, and zero or more non-intersecting, non-nested inner rings.

- *Simple polygon.* Polygon without inner rings; boundary does not intersect itself.

- *Complex polygon.* Polygon with one or more inner rings.

In addition to the above definitions, two commonly used features are pixels and grid cells. A *pixel* is a two-dimensional picture unit that is the smallest nondivisible element of an image. A *grid cell* is a two-dimensional object that represents an element of a regular tessellation of a surface.

The definitions of cartographic objects specified by the NCDCDS are provided here for reference only. In spatial analysis, several terms are defined slightly differently depending on GIS software. For instance, the terms "arc" and "label points" as defined in ESRI's ARC/INFO GIS are different from those of the NCDCDS. The definitions in this book generally follow definitions adopted by manufacturers of GIS software.

Basic Elements of Spatial Information

The analysis of spatial order and spatial association requires the following three elements of spatial information: *location*, *attribute data*, and *topology*. First, the exact location of every spatial feature must be available. In a GIS, the location of a feature is expressed by x and y coordinates on a Cartesian plane.

Second, attribute data provide important information about the properties of the spatial features under consideration. For instance, a map showing the distribution of point features must provide necessary information about each point on the map. A point may represent a well, utility pole, water tank, or any other type of point feature. If the point denotes a well, then other attribute data may be needed to describe well properties such as depth, quality and quantity of water, and ownership.

Third, topology is defined as the spatial relationship between map features. In the case of polygons, you may need to know if a polygon falls inside the boundaries of another polygon, that is, if one polygon is contained by another. For line features, you may need to know if two line segments are directly connected, indirectly connected through other segments, or completely separated without any possible linkage. For point features, you may need to know if a point is closer to a specific location than other similar locations. While location and attribute data are understood by most people, topology is often neglected and less easily understood.

Topological characteristics required for most spatial analyses include the following:

● *Adjacency*. Implies whether two polygon features are adjacent to one another. Adjacency is required when the analysis attempts to determine spatial relationships based on neighborhood information. For instance, the expected price of a residential parcel is often higher than average if it is located next to a park.

● *Containment*. Indicates whether a single feature is contained within the boundaries of a polygon. Containment is important when the spatial relationship between two types of map features is required. For instance, parcels located within a flood zone may be subject to higher insurance rates.

● *Connectivity*. Indicates whether two line segments are connected. Connectivity is especially useful for transportation and routing analyses. The connectivity in a transportation network is important for determining the shortest path from a given address to another.

In addition to the above characteristics, intersection is considered a complicated form of spatial relationship between polygon features. Intersection implies that two polygons share a common area which is a subset of both polygons. While GISs provide map overlay functions for processing polygon intersections, intersection is not considered a required topological characteristic. The common area is more effectively organized as a separate polygon in the database with the associated spatial relationships represented by a combination of adjacency and containment.

GISs are indispensable for spatial analysis because of the ability to integrate all three elements of spatial information in a logically consistent manner. A database management system handling only attribute data is at best useful for aspatial statistical analysis. A computer system capable of handling location and attribute data but not topological elements is suitable for automated cartography, but not spatial analysis. A typical automated cartography system provides mapping functions for the organization and presentation of spatial information. In a spatial analysis, complex spatial relationships among map features can be effectively processed only by using a GIS that provides the functionality to handle all three types of elements.

Required Data Elements

Spatial objects are represented as point, line, or polygon features. Each feature type is associated with a specific set of data requirements.

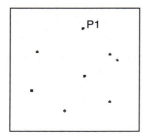 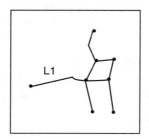 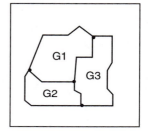

Examples of point, line, and polygon spatial features.

Points are the simplest form of spatial features and require the least amount of data for spatial analysis. In general, the minimal data elements required for representing point features include location and attribute data. Each point must be coded with an explicitly defined position expressed in a coordinate system. In addition, attribute data for variables related to a specific analysis are required. The required data elements for every point feature can be summarized as follows:

$$P_i: (X, Y, Z_1, Z_2, \ldots Z_m)$$

where i is the identification code (ID) of the feature; X and Y stand for the x and y coordinates of the point on the Cartesian plane specified by the x and y axes, respectively; $Z_1, Z_2, \ldots Z_m$ denote values of the point for variables Z_1 through Z_m.

Because point features are zero-dimensional, they do not occupy space; consequently, topological relationships are inherent in their locations. Information about spatial relationships between point features can be obtained indirectly through mathematical operations. For instance, among multiple point features, the nearest point from a given location can be identified by comparing the distance measures between the given location and every other point. As such, it is not necessary to explicitly incorporate topological relationships in data elements of point features.

Line features can be treated as combinations of ordered point features. Every line feature can be decomposed into numerous straight line segments, and each segment is denoted by two points at both ends. In addition, topological elements must be explicitly given in order for a GIS to construct spatial relationships of line features. One fundamental topological element may be the line direction.

Another important topological element could be the connectivity of line features. For instance, you may need to know whether a line is connected to another in order to determine if there is a way to travel from one point to another through the lines. Accordingly, base elements of a line feature include the following:

$$L_j: (P_1, P_2, \dots P_n, Z_1, Z_2, \dots, Z_m, \eta_1, \eta_2, \dots \eta_q)$$

where j is the ID code of the line feature; points P_1 through P_n delineate a specific line in a strict order;

and Z_1 denotes the attribute value of this line for variable Z_1. The direction of the line is indicated by the order of the point sequence. Additional data may be recorded for line feature connectivity. For instance, η_q denotes the ID of q-th line feature directly connected to the line. A more effective method for recording connectivity is discussed in a later section on the arc-node data model.

Polygon features are the most complicated among the three types of spatial features. Polygons are defined by a series of line features delineating boundaries. In addition, because polygons are two-dimensional, each polygon occupies an explicitly defined area. Because polygons are of irregular size and shape, spatial relationships among polygon features are difficult to trace unless a well-designed data model is employed. Two polygons may be completely separated, adjacent to each other, one partially containing the other, or one completely containing the other. In the case of two separate polygons, it is possible to ask whether the two are connected through other polygons.

The required elements of polygon features include the following:

$$G_k : (L_1, \ldots L_n, Z_1, \ldots Z_m, \delta_1, \ldots \delta_r, \varphi_1, \ldots \varphi_s, \phi_1, \ldots \phi_t)$$

where k is the ID code of a specific polygon. A connected sequence of line features, L_i to L_n, delineate the boundaries of the polygon. Z_n denotes the value of the n-th attribute. $\delta_1, \ldots \delta_r$ represent one or more polygons adjacent to polygon k, $\varphi_1, \ldots \varphi_s$ denote one or more polygons that contain polygon k, and $\phi_1, \ldots \phi_t$ denote one or more polygons contained within polygon k.

Additional topological information may be required for more complex polygon configurations. Some of these topological elements may be required for more complex polygon configurations. Some of these topological elements may be simplified or embedded in the line topology. For instance, two polygons are adjacent to each other if they share at least one common border. The common border can be detected from the line features that delineate the polygons.

Data Structures

Spatial data can be organized in either a raster or vector data structure. In the raster structure, spatial features are organized in a regularly spaced coordinate system. Examples of raster-based GIS systems include IDRISI by Clark University and SPANS by INTERA TYDAC Technologies, Inc. Alternatively, in the vector structure, spatial features are defined and organized by combinations of vectors. ESRI's ARC/INFO and PC ARC/INFO and Genasys by Genasys II, Inc. are examples of vector-based GISs. The data structure adopted in a GIS determines how different pieces of information are organized and processed, and thus also determines the properties of the GIS for performing spatial analyses. (See Burrough, 1986 and Clarke, 1995 for a comprehensive comparative discussion of the two data structures.)

Raster Data Structure

Because spatial features are geographically referenced, they can be represented on any map of a known coordinate system. The raster structure requires that each feature be represented in picture units (pixels). In this case, a map is decomposed into pixels, each of which is referenced by its row and column positions. The smallest point feature is represented by a single pixel and has an implied area equivalent to the size of the pixel. The diagram at the right in the next illustration shows that a line feature is represented by a series of connected pixels. Likewise, a polygon feature is represented by a cluster of pixels of the same value.

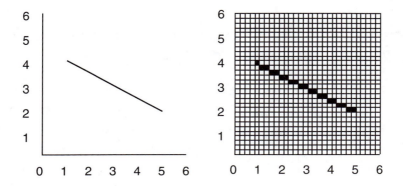

A line can be organized in a vector structure (left) or a raster structure (right).

In general, the raster data structure offers the advantages described below:

● Simple and easy to reference.

● Map overlays can be efficiently processed if the data layers are coded in simple raster structure.

● For spatial modeling, the geographic units defined in the raster structure are consistent in size and shape. As a result, spatial relationships among pixels are constant and easily traceable.

● Generally better representation of continuous surfaces.

● A large amount of spatial data are already in the raster format such as satellite imagery and scanned aerial photos. Consequently, a raster structure database can be constructed by directly importing the raster data from existing sources without the need for laborious conversion.

Disadvantages associated with the raster structure include the following:

● Data redundancy. In most cases, map features do not need to be coded by a dense grid. Vast areas of similar feature value usually exist. For instance, in a simple raster structure a large plain is coded by a huge number of pixels of identical value. Another example is the coding of a surface of relatively constant slope. In the profile of such a slope, the raster structure needs to record a value at a constant distance interval. In brief, data volume tends to be unnecessarily high in raster data structures. Several methods have been developed for converting raster databases into more compact forms. However, every compact form also has corresponding advantages and disadvantages. Whenever a compact form is used to organize raster data, the advantages of the simple raster structure may disappear.

● Topological relationships among spatial features are hard to trace and hard to build. When two map features are referenced by respective row and column numbers, it is difficult to determine how they are related in space.

● Crude raster maps are less aesthetically pleasing than maps drawn with fine lines. Linear features such as roads, rivers, boundaries, and the like are poorly represented by pixels.

● Transformation of spatial data coded in a raster structure tends to be distorted. For instance, a line feature rotated by a specified angle and then rotated back may be different from its original shape.

● For spatial analysis, the most serious disadvantage is that the accuracy in computation and processing of spatial features tends to be lower than desired. For instance, the distance of a straight line can be precisely computed if the coordinates of both ends are known. However, if the segment is recorded in a raster structure, then the length can only be approximated, at best, as seen in the previous illustration.

Vector Data Structure

In the vector data structure, every spatial feature is represented by a set of vectors. In mathematical terms, a vector is specified by a starting point (with given x and y coordinates), a direction (i.e., an angle toward east, west, north, south or some indeterminate direction), and length. A point feature is represented by a "degenerate" vector with both direction and length equal to zero. In this case, the point feature is not associated with a valid measure of area.

A line feature is represented by a sequence of vectors, where each vector represents a straight line segment such as the segment at the left in the previous illustration. Because the width of a vector is not a valid measurement, line features are one-dimensional; only the length of connected vectors is meaningful.

A polygon feature is represented by a series of vectors that form an enclosed area. The area of a polygon is a valid measurement.

Advantages of the vector data structure follow:

● Less data redundancy. Due to organization in a compact format, the vector structure provides an efficient way of organizing large quantities of spatial data.

● Discrete features are represented clearly and continuously. For instance, line drawing is much more efficient.

● Topology of spatial features can be more clearly identified.

● Greater precision in computation of spatial properties and processing of map features.

The most significant disadvantage of the vector data structure is that map overlay, a simple operation in the raster structure, becomes a difficult task. For instance, determining whether a point feature is located inside a polygon can be a straightforward procedure in the raster structure because all you need to know is whether the coordinates of the point (indicated by row and column) are part of the coordinates pertaining to the polygon. The same procedure in the vector structure, known as "point-in-a-polygon," requires complicated computations.

The following example illustrates the difference between the raster and vector data structures in the point-in-a-polygon procedure. The next illustration shows a raster representation of four polygons, labeled A through D. Because the value of every pixel is recorded in the data, it is a simple and straightforward procedure to determine the polygon value of any given point location specified by row and column. For instance, the polygon value of row 7, column 8 is easily identifiable from the file.

The value of a specific point is easily identifiable from a raster database.

Column 8

A	A	A	A	A	A	A	B	B	B
A	A	A	A	A	A	B	B	B	B
A	A	A	A	A	B	B	B	B	B
A	A	A	A	B	B	B	B	B	B
A	A	A	B	B	B	B	B	B	B
A	A	C	C	B	B	B	B	B	B
C	C	C	C	C	D	D	B	B	B
C	C	C	C	C	D	D	D	D	B
C	C	C	C	C	C	D	D	D	D
C	C	C	C	C	C	D	D	D	D

Row 7

The following illustration shows a vector representation of the same map of polygons labeled A through D. In the database, each polygon is represented by a series of x and y coordinates. To determine from the data set which polygon contains the point at 8,4 Cartesian coordinates (equivalent to row 7, column 8) is a difficult task. Each polygon is specified by a series of x and y coordinates, but there is no simple way to determine from the database whether this location is within a polygon.

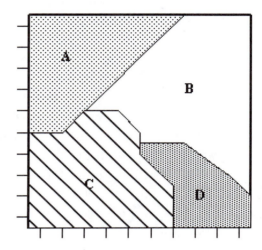

The value of a specific point is difficult to identify in a vector database.

Arc-node Data Model

A data model is the fundamental structure of a database designed to organize diverse components of data for effective retrieval and processing. The data model lays the foundation of a GIS in the sense that the design of the data model determines how the data are organized and processed and consequently sets the strengths and weaknesses of a GIS.

The arc-node data model is an effective vector based data model first adopted by the U.S. Census Bureau for building the geographic base file (GBF) of the 1980 census. Based on this model, streets and other line features in the United States are organized in a file system known as dual independence map encoding (DIME).

In the arc-node data model, arcs form the most basic units on a map. The model's definition of *arc* is different from that of the NCDCDS. In this model, every arc consists of two nodes, a start node and an end node. Between the nodes an arc could have zero or any number of vertices. The shape and length of an arc are determined by the locations of the nodes and vertices. A *node* is different from a vertex in that nodes are topological features, that is, each node is specified by both x and y coordinates and topological significance. A *vertex* (point) is a simpler point feature specified by x and y coordinates without topology.

The following illustration shows an example of a map comprised of four regions represented by polygons (101, 102, 103, and 104). The polygons are delineated by seven arcs (33 to 39). Note that every arc has two nodes (start and end). Arc 33, which represents the simplest form of line features, has no vertex between the nodes, and thus depicts a straight line segment. Other arcs have one or more vertices between nodes to delineate a more complicated shape (e.g., arc 36 has four vertices). An important requirement of the arc-node data model is that either end of an arc is always defined by a node.

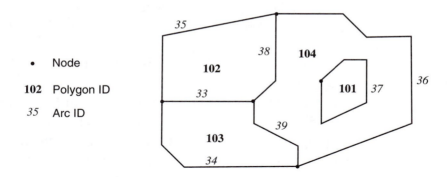

A simple polygon coverage of the arc-node data model.

The direction of an arc is implicitly specified in the order of the nodes, that is, start and end. With the direction known, the two sides (left and right) can be traced. A file containing requisite information about arcs contains the following fields:

● Arc ID

● Node ID of start node

● Node ID of end node

● Polygon ID of the polygon at the right side of the arc

● Polygon ID of the polygon at the left side of the arc

● X,Y coordinates of the start node

● X,Y coordinates of the end node

● X,Y coordinates of all vertices

In addition, a field containing the calculated length of the arc can be included for convenience.

Nodes are topologically meaningful in that they represent the junctions between line features. In the arc-node data model, a point feature can be treated as a degenerate line feature with the start node overlapping the end node and no vertices in between. Each point feature can be represented by one node because start and end nodes are identical. Each polygon feature in the previous illustration, however, is treated as a series of connecting arcs that delineate the boundaries of the polygon. In order to construct the topological relationships among polygon features, every intersection must be defined as a node. The reason for this practice is that crossing two arcs without defining a node at the intersection would make identifying the relationship between two adjacent polygons impossible.

The most important advantage of the arc-node data model is its effectiveness in coding topological relationships. For instance, assume that you want to know whether two polygons are adjacent to one another. A simple way to find the answer is by looking at the list of arcs that define these two polygons. If at least one arc exists which is part of both polygons, then the two polygons are adjacent to one another. In brief, the polygons share an arc.

ESRI's ARC/INFO is based on the arc-node data model and organizes spatial feature data in separate, relational data files. Appearing below are the two most important attribute tables for spatial analysis incorporated in ARC/INFO–the polygon attribute table (PAT) and the arc attribute table (AAT)–for the polygon coverage shown in the previous illustration.

PAT

#-ID	Poly-ID	Perimeter	Area
1	0	8.418	-4.506
2	104	8.596	2.078
3	102	4.296	1.144
4	101	2.233	0.301
5	103	4.325	0.983

AAT

#-ID	Arc-ID	F-node	T-node	L-poly	R-poly	Length
1	38	3	1	3	2	1.151
2	33	4	3	3	5	1.040
3	35	4	1	1	3	2.105
4	37	2	2	2	4	2.233
5	36	1	5	1	2	4.120
6	39	5	3	5	2	1.093
7	34	4	5	5	1	2.193

The PAT lists all polygon features in the coverage. The first record in the PAT, called the "universe polygon," represents the aggregated area extent of all polygons in the coverage. The area is the negative of the total area coverage. Next, the four valid polygons are referenced internally by a #-ID and a user-specified polygon ID (i.e., 101 to 104). The area and perimeter of the polygons are computed when the coverage is generated.

In the AAT, every arc is a single record referenced by both an internal #-ID and a user-specified arc ID (i.e., 33 to 39). The start node (F-node) and end node (T-node) and the polygons on both sides of each arc (L-Poly and R-Poly) provide the necessary topological information about the polygons and arcs. Note that the L-Poly and R-Poly refer to the internal #-ID of polygons in the PAT (e.g., L-Poly 3 in the AAT is equivalent to the polygon with the Poly-ID of 102 in the PAT). The length of every arc is also computed when the coverage is generated.

The three most important elements of topology—adjacency, containment, and connectivity—are embedded in the arc-node data model. To determine whether two polygons are adjacent, the analyst can simply examine the arcs that delineate both polygons to determine if an arc exists that is shared as a common border between polygons. For instance, to determine whether polygons 102 and 103 are adjacent, the analyst can examine the L-Poly and R-Poly in the AAT to identify any arc which has the two polygons on both sides. In the example, arc 33 shows that polygons 102 and 103 are adjacent to one another.

Containment can be examined in a similar manner. To determine whether polygon A is contained by polygon B, the analyst must first select all arcs that delineate polygon A, and then examine the polygon IDs on both sides of the selected arcs. If all selected arcs have polygon A on one side and polygon B on the other side of every arc, then polygon A must be contained by polygon B. For example, because polygon 101 is delineated by an arc (arc 37) for which the other side is always polygon 104, it is clear that polygon 101 is contained within polygon 104. This procedure may be modified to handle more complex situations such as a single large polygon containing multiple adjacent polygons.

Connectivity of arcs can be determined from the attributes of the F- and T-Nodes in the AAT, that is, two arcs are directly connected if both are linked to the same node. Because arcs 38, 33, and 39 share a common node, these three arcs are directly connected. However, arc 37 is not connected to any other arc because no other arc is linked to the only node defining it.

The above example illustrates an important property of the arc-node data model: topological relationships are traceable because the required topological information is incorporated in the data files. In addition, location data organized by the x and y coordinates for every node and vertex in internal files can be retrieved as necessary. Attribute data can be either attached to the AAT or the PAT, or organized in other tables for referencing through the relational database. Because all three required components of spatial data are integrated in this data model, the ARC/INFO GIS provides an effective tool for processing spatial information and conducting complicated spatial analyses.

The arc-node data model, or a modified form, has been adopted by manufacturers of well-developed vector-based GISs such as ESRI and Intergraph. Map-Info (by MapInfo Corporation), a desktop GIS, does not follow this model. As a result, a database constructed in MapInfo cannot support advanced analytical functions that require the processing of information about topological relationships. In essence, the capability of embedding topological relationships in the arc-node data model provides the foundation for vector-based GIS to effectively handle complex spatial analysis tasks.

Summary

This chapter focused on spatial data, which is the foundation of GIS-based spatial analysis. Statistical analyses are usually aspatial in nature and do not require spatial components of the objects under consideration. Spatial analysis requires that objects be geographically referenced, or mappable. In other words, a spatial analysis requires processing information about objects that can be represented on maps.

Spatial features are represented on maps in different forms: points, lines, and polygons. Critical elements of spatial information associated with each object include location, attribute data, and topology. The organization of these elements is different between raster and vector based data structures. In general, location and attribute data are organized in much the same way in both structures, whereas the organization of topology makes a remarkable difference between these data structures. Topology is defined as the spatial relationships between map features.

In spatial analysis, the most important information about topology includes adjacency, containment, and connectivity. A major disadvantage of raster data structure is the difficulty in tracing topological relationships. Among vector data structures, the arc-node data model described in this chapter provides the most efficient means for organizing topological information. Indeed, vector-based GISs that adopt the arc-node data model provide the most effective tools for spatial analysis.

Exercise

1. Construct the arc attribute table (AAT) for the diagram below. If you have access to a vector based GIS such as ARC/INFO or PC ARC/INFO, digitize the diagram and build a line coverage to generate the AAT. Alternatively, you can manually create the AAT based on the discussion of the arc-node data model in this chapter. In manual construction of AATs, the length of each line segment can be roughly estimated because accuracy is not important in this exercise. Then, according to the AAT without referencing the diagram, answer the following questions: (a) Are nodes A and B directly connected? (b) Are nodes A and C directly connected? (c) Find at least two ways to move from node A to node F.

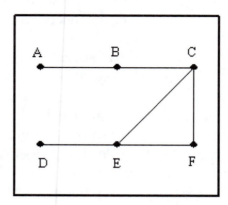

2. Answer the following questions according to the PAT and AAT below. (a) How many polygons and arcs are there in this coverage? (b) Which polygon is contained within another? (c) Which polygons are adjacent to Polygon 102?

Manually construct a simple diagram to show the spatial arrangement of this polygon coverage.

PAT

#-ID	Poly-ID	Perimeter	Area
	0	18	-18
2	110	12	9
3	111	16	8
4	112	4	1

AAT

#-ID	Arc-ID	From-node	To-node	L-poly	R-poly	Length
1	71	1	2	2	1	9
2	72	1	2	3	2	3
3	73	2	1	3	1	9
4	74	3	3	3	4	4

Quantification of Spatial Analysis

The three elements of spatial information—location, attribute, and topology—play different roles in spatial analysis. Typically, the primary research question of a study is focused on attribute data. In a regression analysis, models are built to explain the distribution of a phenomenon (the dependent variable). The dependent variable in turn is defined as an attribute of spatial features rather than by location or topology. For instance, in analyzing the spatial distribution of household income, the geographic unit of the study is household, whereas the theme for the analysis is income, an attribute of household.

However, spatial relationships among geographic features are based on location data. The spatial distribution of households and their interrelationships are recorded in the location data rather than the attribute data. Furthermore, topological relationships among geographic features are implicitly embedded in the location data. Quantification of these elements, which is a prerequisite to any analysis of spatial order and spatial association, forms the central theme of this chapter. Specifically, this chapter deals with issues pertaining to the measurement of both location and attribute data. In quantitative spatial analysis, every element of location and attributes must be measured. The quantification of location allows spatial features to be geographically referenced and mapped, while the quantification of attributes allows for the distribution of spatial relationships to be revealed and analyzed.

Measurement of Location

Every spatial feature takes place at a specific location on the surface of the Earth. In order to represent the feature on a map, the location must be expressed in terms of an established coordinate system. The accuracy of the location depends on the coordinate system employed and the way in which the coordinates of map features are organized. Next, the analysis of spatial relationships usually requires handling different data sources that may be organized using different coordinate systems. Therefore, a GIS suitable for spatial analysis must provide the following important functions dealing with coordinate systems.

● Commonly adopted coordinate systems must be implemented so that data organized in a given coordinate system can be processed in the GIS. When processing location data from different sources, the GIS must have the capability of integrating diverse coordinate systems.

● Because GIS applications frequently involve the use of multiple data sources which organize data in different coordinate systems, the GIS must provide built-in utilities for coordinate conversions. The GIS should at least allow the coordinate system of any spatial data set to be converted back and forth among commonly used systems.

● The GIS must also allow for the user to convert any coordinate system into a user-specific coordinate system. As long as the user is able to explicitly specify the rules governing the desired coordinate system, the GIS must provide suitable functions for coordinate conversion.

Measurement of location is the first step toward spatial analysis. The key issue is how to represent the location of any object in a systematic manner. A prerequisite is the establishment of a coordinate system. Understanding frequently used coordinate systems is important because the coordinate system used in a study determines the quality and properties of location data. The most important system to cartographers and spatial analysts is the geographic grid system, which is widely accepted and referenced by all other coordinate systems.

Geographic Grid System

In the geographic grid system, the location of a place on the surface of the Earth is referenced according to distance from the equator and prime meridian. *Geographic north* (north pole) is defined as the northern end of the rotational axis of the Earth and, likewise, *geographic south* is the southern end of the axis. The equator is the locus of all points defining an imaginary circle on the surface of the Earth midway between geographic north and south.

Fundamentals of the geographic grid system.

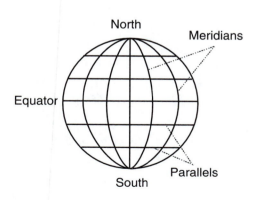

A *parallel* is any circle on the surface of the Earth that is parallel to the equator. Thus, there are an infinite number of parallels on the sphere, and every point on the surface of the Earth falls on exactly one such parallel. A *meridian* is an imaginary circle on the surface of the Earth which bisects the sphere while passing through both geographic north and south. Accordingly, there are also an infinite number of meridians on the surface of the Earth. Meridians and parallels are always perpendicular to each other, that is, they always intersect each other at 90 degrees. The *prime meridian* is arbitrarily defined as the meridian which passes through the British Royal Observatory at Greenwich.

The equator is used as the *base parallel* for expressing location in the north-south direction. The equator has a designated latitude of 0°. The *latitude* of any point location measures the angular distance between that point and the equator, and is defined by the angle between the straight line from the point to the Earth's center, and the straight line from the Earth's center to the point along the same meridian and on the equator. If such a point is located in the northern hemisphere, that is, between the equator and the north pole, the latitude has a sign of north. If the point is in the southern hemisphere, that is, between the equator and the south pole, the sign of the latitude is south.

The prime meridian, designated 0° longitude, is used as the *base meridian* for measuring location in the east-west direction. The eastern hemisphere is to the east of the prime meridian, and the western hemisphere to the west. The longitude of any point is measured by the angular distance between that point and the corresponding point on the same parallel along the prime meridian. The east or west sign of longitude depends on whether the point is located in the eastern or western hemisphere.

Latitude measures the angular distance from the equator.

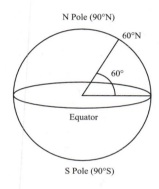

Longitude measures the angular distance from the prime meridian.

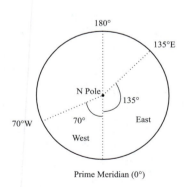

Because longitude and latitude are defined on the basis of the geographic grid system, the measures are derived from a sphere instead of a flat surface. As such, each block delineated by any pair of meridians and parallels always has a surface curvature. The coordinates based on longitude and latitude must be converted into other coordinate systems through map projections. In brief, the geographic grid is projected from its spherical form onto a flat surface for mapping purposes.

The following two interrelated spatial properties of longitude and latitude are important to spatial analysts.

- The distance of 1° latitude is generally regarded as a constant, although in fact the distance varies depending on latitude. However, the distance of 1° longitude varies from latitude to latitude.

- The distance of 1° longitude is different from the distance of 1° latitude, and the difference between the two also varies from place to place. Due to the inconsistency in measurement between longitude and latitude, the measurement of distance based on longitude and latitude is not generally applicable for spatial analysis.

The geographic grid system is a general reference of location on the surface of the Earth. Because this system is defined on a sphere rather than a flat surface, its properties are different from a two-dimensional representation of location. Consequently, for empirical applications location data must be converted to a two-dimensional coordinate system. Three issues are important for establishing the coordinate system: map orientation, map scale, and map projection.

Map Orientation

When defining a coordinate system, the first issue is how to orient the system. Map orientation indicates the direction on a map commonly represented by *north arrows* (arrows pointing to the north). The U.S. Geological Survey's topographic map series typically shows three directions: true north, magnetic north, and grid north. True north is the direction pointing to geographic north (north pole). *Magnetic north* refers to the direction in which compass needles point. *Grid north* is the vertical line pointing to the northern end of the grid system. By convention, the vertical axis of a specific coordinate system follows the same direction as grid north. Because grid north may vary for different parts of the Earth and for different coordinate systems, it is the analyst's responsibility to understand the direction of the coordinate system.

Map Scale

Map scale is defined as the ratio of a unit distance on the map to the true distance of the corresponding unit on the ground. Map scale can be expressed either verbally, mathematically, or graphically.

● *Verbal.* Examples are "One inch represents four hundred feet," or "One inch equals four hundred feet."

● *Mathematical.* These expressions usually take the form of a representative fraction such as 1:4,800 or 1/4,800, which relates one map unit of distance to 4,800 actual distance units on the surface of the Earth. The representative fraction is conventionally called the *RF scale* by cartographers.

● *Graphical.* Typically shown in the form of scale bars on maps.

Because scale is denoted as a fraction, a large value has a smaller denominator. Thus, a large-scale map covers a smaller area in detail while a small-scale map covers a larger area and provides less detailed information. For example, a 1:24,000 scale map is represented at a smaller scale than a 1:12,000 scale map. The features in a 1:12,000 scale map will typically be drawn larger and the entire map will contain more detail than a 1:24,000 scale map.

For spatial analysis, three properties related to map scale are important.

● Map scale determines the *precision level of location data.* The precision of location data is influenced by a number of factors, including the hardware precision or resolution of input and output devices (i.e., the resolution of a digitizing tablet or plotter), accuracy of the source map, design of the database, software resolution (i.e., the definition of a variable in terms of integer or floating point, and, in the case of a floating point, single precision or double precision), and others. All things being equal, the precision of the location data is a function of map scale. In general, a larger map scale implies a higher precision level. For instance, if the U.S. Census Bureau's TIGER/Line file derived from a source map with a scale of 1:24,000 and an estimated ground accuracy of 50 ft, the same level of ground accuracy at a map scale of 1:12,000 would be 25 ft.

● Map scale influences the *measure of spatial statistics*. For instance, the measure of spatial pattern indicated by Moran's *I* coefficient of spatial auto-correlation is a function of map resolution, whereas map resolution is determined by map scale (Chou, 1991). Spatial analyses dealing with complicated spatial relationships may require the use of spatial statistics for modeling and hypothesis testing. The analyst must be aware of the fact that spatial statistics tend to be affected by map scale and must therefore use the statistics cautiously.

● Although the scale is explicitly provided on a standard map, in reality *map scale varies from place to place depending on map projection*. Issues pertaining to map projection will be discussed in the following section. The spatial analyst must be aware of the fact that the variation of scale on different parts of a map may distort certain measures. For instance, the distance between two points near the equator could be quite different from another two points located at a high latitude, even though the distances on the map appear to be similar. This distortion may be negligible for applications based on large-scale maps (i.e., the problem is not significant for analyses that involve only a small area). However, for spatial analyses at the landscape scale, or analyses that involve a very large area, the distortion of measurements could be quite significant and must be taken into consideration.

Map Projections

Spatial analysis requires conversion of features on the Earth's surface to a two-dimensional coordinate system. Map projection is the procedure for transforming the location data from a sphere to a *developable surface*. A developable surface can be represented on a completely flat surface. A sphere is not developable because it can never be pressed into a perfectly flat surface. Even if the sphere is divided into a large number of little pieces, every piece retains the curvature of the original sphere. Therefore, a required process in cartography is to systematically transform location data from the sphere into a developable surface.

Numerous map projections are available, but all distort one or more of the following properties of the Earth's surface: shape, area, distance, and direction. The optimal map projection for a given project depends on property precision requirements, and the size of the study area, among other factors. Each projection has specific transformation procedures and is associated with specific properties. The projections can be classified into four principal methodology categories: azimuthal, cylindrical, conic, and pseudocylindrical.

The next illustration shows an *azimuthal projection* which is characterized by the use of a light source at the center of the sphere to project the *grat-*

icule (meridians and parallels) from the sphere onto a tangent plane. The diagram to the right indicates that the azimuthal projection generally results in a circular shape of the graticule. The location of the light source may vary from projection to projection. For instance, the light source for the *stereographic azimuthal projection* is located at an infinite distance from the Earth.

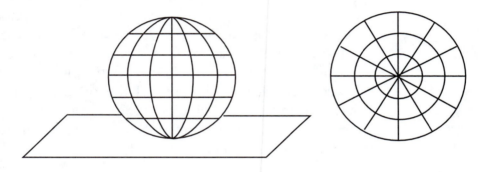

An azimuthal projection transforms the graticule onto a tangent plane.

A *cylindrical projection* transforms the graticule from the sphere onto the surface of a cylinder, where the cylinder either covers the sphere with a tangent circle or cuts through the sphere to form two parallel, intersecting circles. The surface of the cylinder is then developed into a flat surface to define a two-dimensional coordinate system. The following illustration shows an example cylindrical projection.

When developed from the surface of the cylinder, the graticule appears to be rectangular in shape.

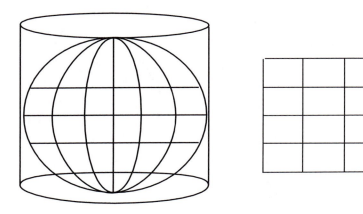

A cylindrical projection results in a rectangular-shaped graticule.

A conic projection either covers the sphere or cuts through the sphere with a cone. The graticule projected onto the surface of the cone is then converted into a two-dimensional representation with a fan-like shape. Other projections such as the so-called *pseudo-cylindrical projections* are not truly cylindrical, but are derived mathematically with certain properties resembling those of a cylindrical projection.

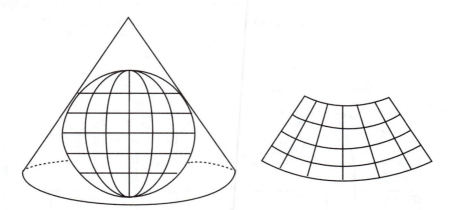

A conic projection generates a fan-shaped graticule.

Map projections can be divided into three spatial property types—equivalence or equal-area projections, conformal projections, and other. In an *equal-area projection*, the area of every map feature is preserved. A *conformal projection* preserves the angular relationships or shapes of map features.

Area and angular relationships are mutually exclusive properties; a projection which preserves both properties is not possible. In other words, an equal-area projection always distorts the angular relationships among map features, whereas a conformal projection always distorts area relationships.

The "other" type of map projection consists of projections that are neither equal-area nor conformal, but are usually some compromise between these two properties. Equivalence and conformity are both global properties; a conformal projection preserves conformity everywhere on the map, while an equal-area projection preserves area everywhere. Some projections may introduce other local properties. For instance, *equidistant projections* preserve distance as a local property that does not apply everywhere on the map. Another example is the *azimuthal equidistant projection*, which preserves distance between the point at the center of the projection and every other point on the map. In this case, equidistance does not apply to any other pair of points.

The following illustration shows California county boundaries represented in the geographic grid system of longitude and latitude. In this case, the x axis represents latitude and the y axis longitude.

*California county
boundaries
represented in the
geographic grid
system.*

    The next diagram illustrates the same map con-
verted into an equal-area projection known as the
sinusoidal projection. Note that the shape of California is quite different from the geographic grid representation. Consequently, the x and y coordinates of
every point delineating the boundaries are different.
Moreover, note that the area of a county is different
when computed from a different coordinate system.
Because the sinusoidal projection is equal-area, the
area of any county computed from the map of this
projection represents the true area of the county.

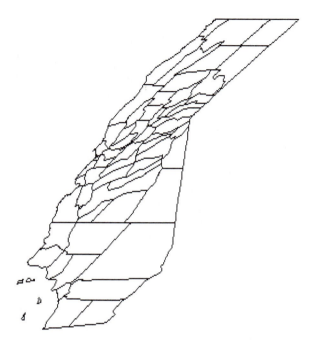

Representation of California county boundaries in the sinusoidal projection.

Coordinate Systems

Analysis of spatial relationships requires consistency of spatial properties in the coordinate system representing the location of map features. For this reason, all methods included in this book are based on a *Cartesian coordinate system.* In the Cartesian coordinate system, every point location is expressed in a two-dimensional plane defined by x and y axes that satisfy two properties. First, the axes are perpendicular to each other. Second, the unit distance is identical on both axes.

The geographic grid coordinate system is not Cartesian for two reasons. First, the geographic grid is based on a spherical surface which is not two-dimensional. Second, although meridians and parallels are perpendicular to each other, the unit distance on meridians is different from that on parallels. In other words, the distance of 1° longitude is not always equal to that of 1° latitude. Consequently, the geographic grid system is not suitable for establishing the coordinate system for spatial analysis. Among available coordinate systems, the UTM system is the most widely adopted by GIS users in the United States.

UTM System

The UTM coordinate system is based on the Universal Transverse Mercator projection. The U.S. Geological Survey's 7.5-minute series of topographic maps incorporates the UTM coordinate system along with the Universal Polar Stereographic (UPS) system.

In the Mercator projection—historically, the most commonly used projection for navigation—, the standard grid line is along the equator. In contrast, the standard grid line in the transverse Mercator is along a meridian. Because distortion increases with distance from the standard meridian, a transverse Mercator projection is useful only for a relatively narrow zone adjacent to the standard meridian. For this reason, the entire surface of the Earth must be divided into a number of relatively narrow, north-south zones. Coordinates can be represented with minimal distortion within each zone.

The UTM system is organized in such a way that each zone covers an east-west extent of 6° longitude. Thus, in the UTM coordinate system, the entire surface of the Earth is covered by 60 zones. The scale is preserved along the two meridians slightly to the east and west of the central meridian where the scale is set at 0.996 of those lines. The next illustration shows the structure of the UTM coordinate system.

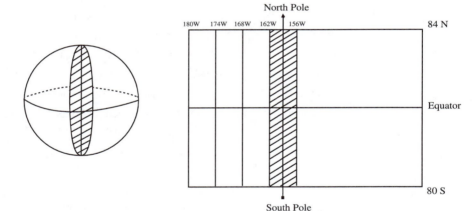

UTM coordinate system.

In the Transverse Mercator projection, the cylinder is rotated 90° with the tangent along a meridian instead of the equator.

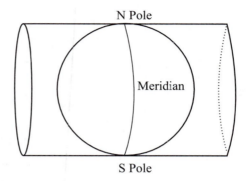

The 60 UTM zones start at the International Date Line (IDL), or 180° longitude, and extend at an increment of 6° longitude to the east. Accordingly, the first zone (Zone 1) covers the range between 180° W and 174° W longitudes. Zone 2 starts at 174° W and extends to 168° W, and so forth. The UTM coordinates are meaningful only within a zone, and the coordinates of different zones are independently referenced. In other words, the coordinates in Zone 11 are not meaningful when compared with those in Zone 12.

Easting is the measure of east-west distance from the center line of a zone, equivalent to the coordinate of the horizontal axis (x axis) in a Cartesian coordinate system. The center line is assigned a standard easting measure of 500,000 m. This value was selected so that every point within any zone has a positive easting value. Thus, an easting greater than 500,000 m indicates that the location is to the east of the center line in the corresponding zone.

An easting value less than 500,000 m represents a location to the west of the center line. In all instances, the easting value is positive and less than 1,000,000 m. Because the unit distance of longitude varies by latitude, the east-west extent is a variable within the same zone. In other words, the actual horizontal extent of a zone is widest along the equator, but becomes narrower as one moves toward either pole.

Northing is the measure of north-south extent, equivalent to the y (vertical) axis coordinate in a Cartesian system. In the northern hemisphere, the equator is designated as the 0 m line. All locations in the northern hemisphere have northing values depending on distance from the equator. In the southern hemisphere, the south pole is the reference point with a northing of 0 m, and the equator has a northing of 10,000,000 m.

The UTM coordinate system is Cartesian in nature because meters are its standard measure. In brief, one unit distance in x is equivalent to one unit distance in y, and the easting and northing measures are vertical to one another. However, a major limitation of the UTM coordinate system is that it is not directly applicable for areas that cross zones. In other words, if a study area happens to be located where two neighboring zones cross (e.g., the eastern half of California is in Zone 11 while western California is in Zone 12), the UTM coordinates must be converted into another coordinate system for analysis.

Because most GISs have built-in conversion functions for major projections, and the geographic grid system is the most commonly used to represent the Earth's surface, the analyst may convert the UTM coordinates of two neighboring zones into latitude and longitude of the geographic grid system. The entire map area must then be converted into another coordinate system for analysis because the geographic grid system is not Cartesian.

Another property especially important for spatial analysis at the landscape scale is that UTM zones are organized in a rectangular shape, yet in reality, they are not rectangular. This is because the east-west extent is a variable—the span of 1° longitude varies from place to place. In general, the UTM system is only valid for area coverage within the range of 84 degrees North and 84 degrees South. Beyond this range, the Universal Polar Stereographic projection is used instead. The coordinates for places crossing the boundaries cannot be mixed.

State Plane Coordinate System

Maps used in most U.S. municipal applications adopt the state plane coordinate system as an historical legacy of the country's land survey system. The system is not appropriate for regional applications due to the fact that different states employ different projections. States of east-west extent are based on the transverse Mercator projection, whereas states of north-south extent are based on the Lambert conformal conic projection. Each state is further divided into subdivisions to minimize distortion.

An advantage of the state plane coordinate system is that its basic unit is the foot, which is more acceptable in the United States than the standard unit (meter) of the UTM system. The acceptability of the foot as a standard unit of distance measurement is based on a long tradition of using the English system, which is based on the inch.

However, the English system does not use a consistent number base: the inch is divided into arbitrary fractions and the foot is comprised of 12 inches, while the yard is three feet and the mile is 5,280 feet. The metric system is more consistent (i.e., the standard unit of distance, the meter, is comprised of 100 centimeters).

Measurement inconsistency makes the state plane coordinate system somewhat problematic for spatial analysis at a regional scale, not only because conversion factors are required whenever conversion between units is needed, but coordinate system translation from one location to another involves cumbersome procedures, especially when different locations are registered to coordinate systems of different map projections.

Basic Measurements of Spatial Features

With an established coordinate system, every point feature can be represented by a pair of x and y coordinates. For the sake of convenience, in this section the P_i expression represents the i-th point feature in a coverage, and the location of this point is denoted by (x_i, y_i).

Line features are represented by an ordered sequence of points. In digital cartography, curvilinear objects are decomposed into sections of straight line segments. The quality of the representation of complicated curves can be improved by increasing the spatial resolution of data points, that is, increasing the number of straight line segments to approximate curvilinear features. As a straight line, each segment is denoted by two end points. Thus, any line feature L_i can be represented by a series of P_js such as L_i $(P_1, P_2, P_3,...P_n)$, where n denotes the number of points that delineate the line L_i.

Area features can be represented in the same way that curvilinear features are approximated by straight line segments. In this case, a circle is not considered a curve, but instead is a polygon comprised of many small lines of uniform length that form an area of circular shape. Accordingly, the polygon G_i can be represented by a sequence of connected lines such as G_i $(L_1, L_2, L_3, ... L_m)$, where m represents the number of connected lines that delineate the polygon.

Once spatial objects are defined, basic geometric measurements can be derived. The distance between two points on a Cartesian plane is evaluated by *Euclidean distance*. The following formula is commonly used to evaluate the length of a straight line segment.

$$D_{ij} = \sqrt{(X_i - X_j)^2 + (Y_i - Y_j)^2}$$

The following illustration shows a simple example of a line segment specified by two end points at (1, 4) and (4, 2). The Euclidean distance as computed by the above formula is approximately 3.61. Accordingly, any line feature can be decomposed into one or more straight line segments, and the length of the feature is the sum total of the Euclidean distances of all segments.

Straight line segment specified by the coordinates of two end points.

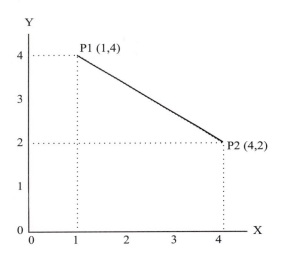

The Euclidean distance measures the straight line distance between two points on a plane. In reality, the actual measure of distance may require modification. For instance, the *Manhattan distance*, which is also known as the *city block distance*, evaluates the distance based on a grid system similar to city blocks. The Manhattan distance formula follows.

$$D_{ij} = \left|(X_i - X_j)\right| + \left|(Y_i - Y_j)\right|$$

The area of any area feature represented as a polygon can be computed by constructing a *trapezoid* from every line segment delineating the polygon, and then systematically aggregating the trapezoid areas. The next illustration provides an example of how this is done. The polygon is delineated by seven straight line segments where each segment is specified by two end points (nodes or vertices). Because the end node of every segment overlaps the start node of the next segment, there are seven nodes defining the seven segments. Two vertical lines are drawn for each segment from both ends of the segment. A trapezoid is defined by the two vertical lines, and the corresponding section on the horizontal axis bounded by the two vertical lines.

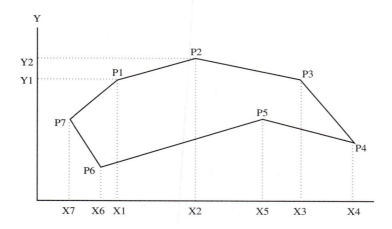

Computation of polygon area based on the trapezoid method.

The area of the trapezoid is computed by the following formula:

$$A_{ij} = \frac{(Y_j + Y_i)(X_j - X_i)}{2}$$

where A_{ij} denotes the area of the trapezoid defined by the two points i and j.

Using the above formula, and assuming that all line segments of the polygon are organized in a clockwise direction, the area of the polygon can be computed by summing all trapezoids in the clockwise direction. Note that the computation specified above assigns positive values to trapezoids defined by the edges of P7-P1, P1-P2, P2-P3, and P3-P4. The computed values for trapezoids defined by P4-P5, P5-P6, and P6-P7 are negative. Thus, the area of the polygon is exactly the computed sum total of trapezoid areas.

A similar method for computing the polygon area is by drawing triangles from both ends of each segment to the origin of the coordinate system, and then computing the sum of all triangle areas. The result of the triangle method is identical to that of the trapezoid method.

Attribute Data Measurement

Both digital mapping and quantitative analysis of spatial phenomena require that spatial objects (or geographic units) be quantified and evaluated based on a clearly defined measurement. The measurement of attribute data is dependent on the nature of the variable under consideration and data availability. Variables are classified into four measurement levels: nominal, ordinal, interval, and ratio.

Measurement Levels

Nominal

Nominal level measurements are qualitative and categorical. Examples of nominal level variables are country of origin, hair color, place of birth, and the electoral district in which voters reside in U.S. states. A nominal value serves merely as a label or name, and no assumption of ordering or distances between categories is made. For instance, a geographic unit whose label begins with "A" does not imply a better/worse or larger/smaller condition than another that begins with "B."

A nominal level measurement can be alphabetical or numerical. When a number is assigned to a geographical area (as in the case of U.S. electoral districts) or any other phenomenon, it is used merely as a symbol. The properties of real numbers, such as being able to add and multiply numbers, cannot be transferred to these numeric categories.

Ordinal

Ordinal level measurements are meaningful in terms of rank order in that each category has a unique position relative to the other categories. For instance, individuals or households may be rank ordered by social class based on occupation or status as follows:

working, middle, and upper. In this case, the middle class is understood to rank higher than the working class and lower than the upper class.

Although ordinal level measurements can be coded alphabetically or numerically, the degree of difference (distance) between one category and another is not known. The characteristic of ordering is the sole mathematical property of this level. The use of numeric values as symbols for category names does not imply that any other properties of the real number system can be used to summarize relationships of ordinal level variables.

Interval

Interval level measurements are meaningful in terms of ordering and distance between categories. The distance between categories is in fixed and equal units. A typical example of interval level measurement is temperature (Fahrenheit or Centigrade). The difference between 50° C and 49° C is equal to the difference between 49° C and 48° C. However, the temperature at 50° C is not twice as warm as 25° C.

Next, in interval level measurements, there is no meaningful starting point. For instance, 0° C is arbitrarily defined by the freezing point of water under specific conditions, and does not imply the absence of heat.

In brief, interval level measurement permits examination of the differences between phenomena, but not their proportionate magnitude.

Ratio

The ratio level of measurement has all the properties of the interval level, with the additional property that the zero point is inherently defined by the measurement scheme. For instance, when measuring physical distance, a zero distance is naturally defined as the absence of any distance between two objects. This property of a fixed and given zero point means that ratio comparisons can be made. Thus, it is meaningful to say that 10 km is twice as far or distant as 5 km.

Additional examples of ratio measurements are weight and income measured in currency units. Zero weight is naturally defined as the absence of weight. It is meaningful to say that 10 metric tons is twice as heavy as 5 metric tons. Zero income is naturally defined as the absence of income or zero dollars (or pesos, yen, and so on). It is meaningful to say that $100,000 is twice as much income or money than $50,000.

Central Tendency and Dispersion

Central tendency and dispersion are the most important descriptive statistics for understanding the distribution of a phenomenon (variable). In statistical terms, *central tendency* shows the trend in the distribution, while *dispersion* shows the extent of dispersion about the central tendency.

For example, assume an income distribution comparison between two ethnic groups. The average (mean) is a central tendency statistic that may be used as an indicator of whether one group tends to make more money than the other. The level of dispersion indicates the significance of the mean. Assume that in Z City the average household income of both African American and Hispanic residents is $25,000 per year. Dispersion analysis shows a small number of extremely wealthy households among the African Americans, while household income is more evenly distributed within the Hispanic population.

For variables measured at the nominal scale, central tendency is evaluated by the *mode*, the class of highest frequency in the distribution. For instance, assume that Y City is comprised of Caucasians (30%), Hispanics (60%), and Asian Americans (10%). The mode of the city's ethnic population distribution is Hispanic, or the largest population group. In this case, the mode represents the central tendency in the sense that if a city resident was randomly selected, the individual's most likely ethnicity would be Hispanic. The frequency of the mode group is denoted as f_{mode}.

The dispersion of nominal scale measurements is evaluated by the *variation ratio*, such that

$$v = 1 - (f_{mode}/N)$$

where v is the variation ratio, and N is the total population of all classes. A smaller value of variation ratio indicates a more concentrated case, and the mode is a better indicator of the trend. Thus, if the largest ethnic group comprises 80% of the entire population, the variation ratio is equal to .2, compared to another case in which the dominant group represents only 40% of the population with a variation ratio of .6.

The central tendency for ordinal scale measurements is typically evaluated by the *median*, which is the case in the middle of the distribution (i.e., the same number of cases are above and below the median). Because the variable is measured on an ordinal scale, all cases can be hierarchically arranged in either ascending or descending order. The median is the case which falls exactly in the middle of the array. For instance, if 15 adults are arranged in ascending order of social class, the eighth individual is the median of this distribution. The median social class implies that the adults who are higher in the ranking are in the same or higher social class, and that the people who are lower in the ranking are in the same or lower social class.

The dispersion of ordinal scale variables is evaluated by decile range, which is the difference in the value of the study variable between the case of the top tenth percentile and the case of the bottom tenth percentile. For example, assume that a questionnaire survey was conducted in which respondents ranked 50 cities in terms of living (residential) preference. The preference level is based on a scale of 1 to 10 where 10 represents highest preference. If the fifth most desirable city (the top decile) is ranked 8 while the forty-fifth city (the bottom decile) is ranked 3, then the decile range is equal to 5 (the difference between 8 and 3). In contrast, if survey results showed that the fifth desirable city ranked 6 while the forty-fifth city is ranked 4, then the decile range is only 2. This example illustrates that a smaller decile range represents a less dispersed distribution.

Central tendency of both the interval and ratio scale measurements is evaluated by the arithmetic mean, such that

$$\bar{Z} = \frac{\sum Z_i}{N}$$

where Z_i is the i-th case in the distribution, $\sum Z_i$ is the sum of the Z values of all cases in the distribution, and N is the total number of cases.

Dispersion about the mean is evaluated by either variance or standard deviation. The variance is defined as follows:

$$\sigma^2 = \frac{\sum (Z_i - \bar{Z})^2}{N}$$

where notations are identical to those previously defined. The standard deviation is the square root of the variance. The larger the standard deviation, the greater the extent of dispersion, while a smaller standard deviation indicates a more concentrated distribution.

In many cases, the mean and standard deviation of certain measurements may not be appropriate for evaluating spatial phenomena when spatial weights are required. For instance, in a distribution where geographic units vary greatly in size, the mean may not be a good indicator of central tendency. This is because the unweighted mean treats every geographic unit as identical regardless of its size. In such cases, the central tendency can be better represented by the *area-adjusted mean* computed as follows:

$$\bar{Z} = \frac{\sum Z_i A_i}{\sum A_i}$$

where A_i denotes the area of the i-th geographic unit. The area-adjusted variance is computed as follows:

$$\sigma^2 = \frac{\sum A_i (Z_i - \bar{Z})^2}{\sum A_i}$$

Central tendency and dispersion are the most basic statistics for describing the distribution of any spatial phenomenon. They can be used to compare spatial patterns in the distribution of different features in order to differentiate variables that affect each feature. Descriptive statistics can also be used to evaluate changes in the distribution of a phenomenon to better understand the processes underlying the spatial distribution.

The following diagram illustrates the spatial distributions of three bird species with different central tendency and dispersion. The distributions are of identical size and geographical configuration, that is, each distribution is organized in a 5 by 5 grid. A shaded cell represents the observation of a bird species while a blank cell denotes the absence of that species. (See Chou and Soret, 1996, for an empirical spatial analysis of bird distributions.)

Central tendency shows the average location of the species. Species A appears to be clustered in the upper left corner. The computed mean of x and y coordinates is (2, 2) for this species. Both species B and C have a computed mean of (3, 3). Thus, species A has a different central tendency than either B or C, whereas species B and C share the same central tendency in the middle of the space.

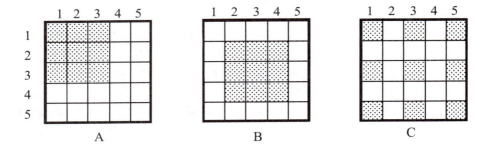

Three spatial distributions that differ in central tendency and dispersion.

Dispersion of the spatial distribution is evaluated by standard deviation. Species A and B, while differing in central tendency, share the same standard deviations of x and y coordinates, 0.83 and 0.83. In other words, the way these two species are dispersed about their corresponding means is identical. Species C, although sharing the same central tendency with species B, has a wider, more dispersed pattern. The computed standard deviations of x and y coordinates for this species are 1.63 and 1.63, indicating that the dispersion is about twice that of either species A or B. In other words, a larger value of dispersion implies a distribution less concentrated around the central tendency.

Summary

This chapter explained how the location and attributes of map features are quantified for spatial analysis. Because maps are two-dimensional representations of reality, the location of any object is denoted by a pair of x and y coordinates. However, because translation of location on a sphere into a two-dimensional surface is always subject to distortion during the process of projection, spatial analysts must be aware of the properties of the projection method employed in the original data source.

Coordinates also vary from one coordinate system to another. In the United States, the UTM and the state plane coordinate system are the most widely adopted coordinate systems. A broad knowledge of map projections and their properties is essential to spatial analysts who must deal with issues pertaining to location data. When performing spatial analysis, you must at least be able to handle the conversion between longitude and latitude in the geographic grid system and UTM coordinates. Once a coordinate system is chosen, basic measurements of spatial features, including the length of line features and the area of polygons, can be derived mathematically and programmed in a GIS.

In addition to measurements of location data, spatial analyses require the processing of attribute data. Attributes are measured at different scales—nominal, ordinal, interval, and ratio. Central tendency and dispersion statistics must be based on the measurement scale of the variable in question. The central tendency and dispersion of nominal variables are evaluated by mode and variation ratio, by median and decile range for ordinal variables, and by mean and standard deviation for interval and ratio variables. These properties characterize the spatial pattern in the distribution of any phenomenon in a spatial analysis.

Exercise

1. Convert the location Lat. 45 N, Long. 171 W into UTM coordinates, assuming that the Earth is a true sphere.

2. Describe the UTM coordinates for 496,000 m easting, 20,000 m northing, Zone 5, using a combination of longitude, latitude, and meters.

3. Digitize the polygon depicted below either manually or using a GIS. First, find the x and y coordinates of each node defining the polygon. Second, according to the coordinates, compute the length of each line segment and the perimeter of the polygon. The perimeter of the polygon is equal to the total length of all segments delineating the polygon. Third, using the trapezoid method discussed above, calculate the area of the polygon from the coordinates.

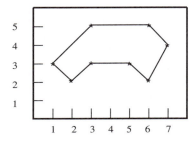

4. Given the attribute information of the polygon coverage below, compute the central tendency and dispersion of each variable.

Poly_ID	LandUse	Intensity	LandValue	Area
1	A	H-	1.0	2.0
2	B	H-	1.5	1.0
3	B	L+	2.0	2.0
4	D	M	1.0	3.0
5	A	M+	0.5	2.0
6	C	H-	2.5	2.0
7	C	H	1.5	3.0
8	B	M-	2.0	2.0
9	B	L+	1.5	1.0
10	B	H	1.0	2.0

Single Layer Operations

Manipulation of spatial information is a critical GIS function because it configures the data to a format appropriate for map presentation and spatial analysis. GISs can be classified into two major data manipulation categories: single layer and multiple layer operations. Single layer operations refer to the procedures that apply to only one data layer at a time, while multiple layer operations operate on multiple layers simultaneously. In empirical applications, these two types of operations are almost inseparable, and most GIS applications require manipulation of every single layer prior to processing multiple layers. For the sake of convenience only, the two types of GIS operations are treated separately in this book.

Single layer operations are also called *horizontal operations* because they deal with only a single data layer and processes only alter data horizontally on a layer. These operations provide the most fundamental tools of data preparation for spatial analysis. More specific and advanced methods of spatial analysis to be discussed in later chapters rely on a GIS to provide the capability of such fundamental operations.

In layer oriented, vector based GIS, each layer contains a single feature type. For instance, all point features may be organized in a single layer, and are not mixed with line or polygon features. In effect, single layer operations deal with only one feature type at a time. Problems that involve multiple feature types (e.g., relationships between point features and polygon features) are treated as multiple layer operations.

This chapter is divided into three sections according to the nature of single layer operations: feature manipulation, feature selection, and feature classification. Issues related to the manipulation of spatial features include boundary operations and proximity analysis. Feature selection involves the identification of features either through graphical manipulations or logical expressions, as well as the description of the characteristics of the identified features. Feature classification involves the classification of features into a suitable number of categories for statistical analysis.

Feature Manipulation

Operational procedures for manipulating map features on a single layer include boundary alteration and proximity analysis. Boundary operations generally manipulate the location data related to map features, and as a result, the boundaries that delineate spatial features are altered. In some cases the altered boundaries define new objects. Proximity analysis consists of operations that generate new polygon features based on distance from selected map features.

Boundary Operations

Common procedures for manipulating feature boundaries include clipping, erasing, updating, splitting, and dissolving. For convenience, in the following discussion the command terminology of ESRI's ARC/INFO GIS is adopted. The command names employed for the same procedures in other GISs may be different.

CLIP

The clipping procedure, CLIP, creates a new coverage which consists of a portion of the original map. All map features within a specific set of boundaries are selected and saved to the new coverage. In short, the new coverage is a subset of the original map, where the subset is defined geographically. This operation is commonly used when only a portion of the map area is to be employed for analysis. For instance, if the transportation network of a city is to be analyzed from a map which covers the entire county, the clipping procedure can be used to extract only those street segments that fall inside city boundaries. In the following illustration the street map of a larger region shown at the left is clipped by the boundaries of a smaller region (the CLIP coverage) to create the new output coverage containing only a subset of the original data.

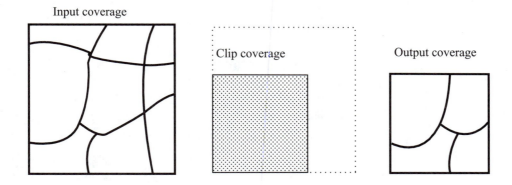

The street map on the left is clipped to produce the map of a smaller area.

The next illustration further explains how the clipping operation works. The input coverage at the left shows five subdivisions labeled 1 to 5. The clip coverage in the middle delineates the area to be included in the resultant coverage. In this case, only the outermost boundary indicated by the outer circle is used to delineate the clip area. The output coverage results from applying the clipping operation. The original input coverage is now confined to a smaller area and the boundaries of the original area units are altered.

Input coverage

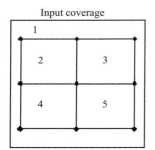

Clip coverage

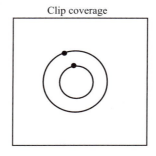

Output coverage

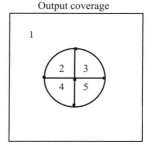

The input coverage is clipped by the clip coverage, resulting in the output coverage.

ERASE

The erasing procedure is the opposite of the clipping procedure. While CLIP copies a portion of the original map to create a new coverage, ERASE removes the portion specified by an intermediate coverage (the ERASE coverage) and retains the remaining features of the original map. In the output coverage, all map features originally located within the specified boundaries are removed. The following illustration shows that map features within the specified boundaries are erased from the original map.

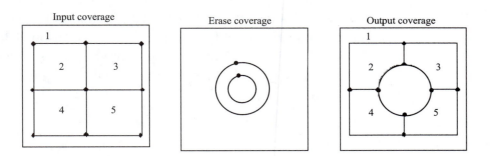

The portion of the input coverage delineated by the erase coverage is removed, producing the output coverage.

ERASE is useful when a map feature subset is to be analyzed, and the features to be excluded from the analysis can be geographically identified and removed. This procedure is also useful when major errors are detected for a particular geographic area. Correcting the database does not require an edit of the entire coverage. A more effective method is to erase the area containing errors and keep the rest of the coverage. The area to be erased may be of regular or irregular size and shape.

UPDATE

Updating replaces spatial data in certain subdivisions on a map with a new or edited coverage. The next illustration shows that the original coverage (input coverage) contains four rectangular units. In this example, the straight lines defining the units are not correct and require editing. Because the input coverage covers a wide area, and it is more efficient to edit the small portion of the original map rather than the entire map area, the "update" coverage is created to replace the corresponding portion in the input coverage. The result of applying the update operation to the input coverage is the output coverage; spatial features are corrected by the update coverage. In other words, UPDATE is an efficient tool for editing a small portion of a large database.

The central portion of the input coverage is updated to produce the output coverage.

SPLIT

Splitting creates subdivision boundaries, dividing the map area into separate coverages. This procedure is useful when a large database must be geographically divided into subdivisions. For instance, the freeway map of a state may be divided into multiple data sets defined by counties. The following diagram illustrates the splitting operation. The upper left corner is the original map which covers a wide area. For efficient processing of the spatial data, a map showing the boundaries of subdivisions (also known as tiles) shown at the lower left is used to divide the original map into four subdivisions. The new coverage of tile

3 is shown in the upper right diagram. The lower right diagram shows the output map of tile 4.

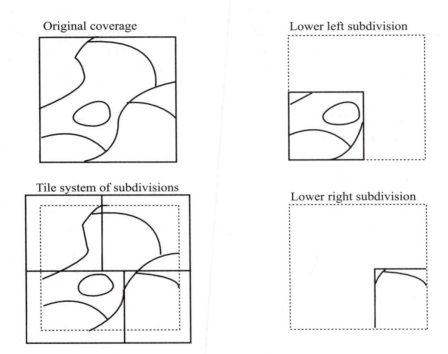

The original coverage is divided into four subdivisions through the split procedure.

APPEND/MAPJOIN

APPEND and MAPJOIN are similar procedures used to place separate, adjacent maps into a single map. These two functions work slightly differently in terms of how the database is reorganized, but the main difference is that APPEND does not rebuild topology or clip the output. Both operations are the reverse of the splitting procedure described above.

The next figure shows a typical example of the procedure for appending adjacent maps. The four soil maps at the left are adjacent to one another, yet the data are organized into separate coverages. When the APPEND procedure is applied, the four maps are merged into the map at the right. It is apparent that additional procedures are required to clean up unwanted boundaries in the output coverage. The dissolving tool solves this problem.

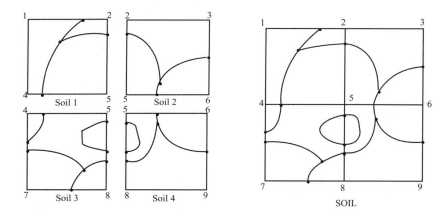

Four adjacent soil maps are appended into the map on the right through the MAPJOIN procedure.

DISSOLVE

The dissolving operation is typically used to eliminate unwanted boundaries between map features after adjacent maps have been appended. DISSOLVE can also be used to rename nodes between lines with identical attribute values. In general, this procedure checks every line segment to determine if both sides of the segment are identical in terms of a user-specified attribute item. For instance, if a segment separates two polygons of the same soil type, as in the previous illustration, then the segment is removed and the two adjacent soil polygons are combined.

The next diagram illustrates the dissolving procedure. The diagram at the left is a map of seven polygons. The polygons are labeled A, B, or C according to the surface type (e.g., vegetation or soil). The dissolving operation examines all segments and eliminates those that separate polygons of the same surface type. Two lines that delineate three polygons of type A are removed, as is the line on the right separating two polygons of type B. After the dissolving procedure is applied, the result is a cleaner output coverage comprised of only four polygons.

Original coverage

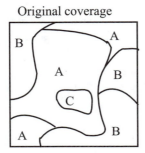

Dissolved coverage

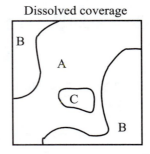

DISSOLVE removes line segments that separate polygons of the same type.

For spatial analysis, the dissolving operation is frequently used in model building. In this case, a model is constructed to describe a spatial pattern and the estimated coefficients are then calibrated to the spatial features. Each feature is then assigned a code according to the estimated coefficients of the model. Boundaries between adjacent features of the same category are then dissolved to generate a new spatial pattern. The use of dissolving operations in spatial modeling will be discussed further in a subsequent chapter.

ELIMINATE

The eliminate procedure is commonly applied in situations where unwanted lines create numerous little, *sliver polygons* due to data errors. Sliver polygon problems are typically the result of map overlay. When two polygon layers are overlaid, the inconsistency in digitization of the two coverages tends to produce numerous sliver polygons. The coverage at the left in the next diagram is a typical result of overlaying two coverages. The ELIMINATE procedure can be applied to remove the lines or longer arcs of the sliver polygons to create a cleaner coverage. (See coverage at the right of the illustration.)

Original coverage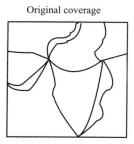

After ELIMINATE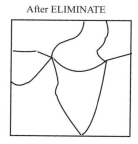

Eliminate procedure selectively removes sliver polygons.

Proximity Analysis

Proximity analysis is based on the distance derived from certain selected features. Area expansion of features, commonly known as *buffer operations* in GISs, are used for several purposes. A buffer operation can be applied to any feature type, including points, lines, and polygons. Because the procedure expands area, it always results in polygon features. The next three illustrations show the buffer polygons generated from a point, line, and polygon coverage, respectively.

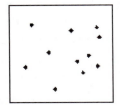 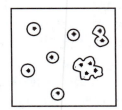 

Buffer zones (middle) are generated from a point coverage (left) resulting in a polygon coverage (right).

Buffer zones generated from a line coverage (left) define a polygon coverage (right).

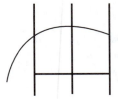 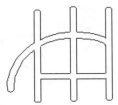

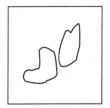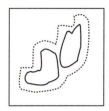

The buffer operation creates an expanded polygon (right) from two separate polygons (left).

In a buffer operation, buffer zone width can be specified either as a constant (i.e., the same buffer width is uniformly applied to all features), or as a variable depending on a specific attribute. For instance, in a study of air pollution, the area extent of possible contamination from pollution sources is a function of the degree of pollutant concentration at each point feature. In this case, the buffer width of potential contamination is a variable depending on pollutant concentration at different contamination sites.

A useful buffer operation for spatial analysis is the generation of *equal distance zones* from selected features. The next figure shows a typical example in which a series of buffer zones are generated, one after another. This method is useful for examining the relationship between the distribution of a spatial phenomenon and proximity to a set of specific features.

For instance, assume you want to study the correlation between property values and the distance from major arterial roadways. For this purpose, the distance between every parcel and the nearest arterial must be estimated. In the current example, the solid lines in the illustration could represent main arterials from which equal interval buffer zones, represented by dashed lines, are generated. Once buffer zones are available, property sale prices in the map area can be correlated with the distance from the main arterials. With this information, you can identify the relationship between land value and distance from arterials. Buffer zones generated in this way are also known as *proximity zones.*

Broken lines represent buffer zones extending from either arterial roadways or rivers (solid lines).

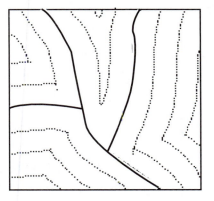

Likewise, the above figure may represent buffer zones (in dashed lines) extended from rivers (solid lines) at a constant interval. In this case, each buffer line is of equal distance (also known as an *isoline*) from one of the original line features.

Another type of proximity analysis involves the use of *Thiessen polygons*, also known as *proximal polygons*. Thiessen polygons are generated from a set of point features, and derived in such a way that each polygon represents the catchment area of a point (i.e., the area inside the polygon is closer to that point than to any other point). The next illustration shows Thiessen polygons generated from a point coverage. In this example, the points may represent major retail establishments competing for business. The generated Thiessen polygons show the potential territory of each establishment based on distance.

Thiessen polygons are created from point features.

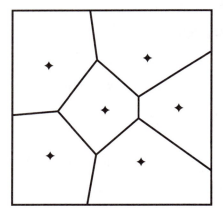

Thiessen polygons are useful for dividing a surface into subdivisions, where the boundaries between two polygons are drawn along lines of equal influence or equal dominance. The previous illustration presents a typical example of Thiessen polygons, created by delineating line segments that bisect each pair of adjacent point features representing competing businesses. Every polygon represents the area of dominance from the point at the center, or the territory of a retail establishment.

Proximity analysis can be extended to multiple layer operations in which the distance between features organized in different data layers is analyzed. Proximity analysis applied to multiple layers is discussed in Chapter 5.

Feature Identification and Selection

GIS horizontal operations are also used to identify spatial features in a map or database. The former can be accomplished through interactive graphical operations, and the latter through logical operations with the database. Interactive graphical operations allow the user to identify features on a map directly from the computer screen. For instance, on a map of point entities, you could move the cursor to a specific location and immediately obtain detailed information about that entity. On a map of polygon features, a GIS allows you to move the cursor anywhere inside a polygon and obtain information on any attribute of the polygon.

The next illustration shows a typical example of interactive, graphical identification of polygon features. As the analyst moves the cursor to any polygon in the land parcel map, a table is displayed which lists the attributes of the selected polygon.

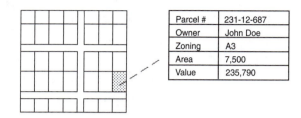

Viewing the properties of a map feature.

The following diagram illustrates an example in which a contour topographical map is displayed on the screen, and the elevation at the cursor location is identified as the cursor is moved around the map area. Alternatively, featuers may be identified through queries or logical operations from the database.

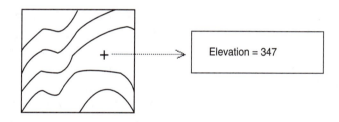

Moving the cursor on the map allows the user to identify the elevation of any point location.

A closely related function is feature selection. In spatial analysis, selecting features from a large database that meet certain criteria is a common operation. Similar to feature identification, the selection of desired features can be obtained by either spatial criteria or attribute conditions. The criteria for selection can be specified in different ways by the user.

For instance, all parcels involved in transactions over the past three years with a sale price greater than $200,000 can be identified from the database and then highlighted on the map.

Logical operations may be conditioned by spatial constraints. For instance, all properties within 500 ft of freeways can be identified through the buffer operations discussed in the preceding section. Likewise, geographic conditions may be employed to select features. Another example is selecting all residential properties within a specific census tract.

Feature selection by spatial criteria can be performed in several ways, described below.

● Directly select features from the map. For instance, the user can move the cursor to pick every desired feature graphically.

● Delineate area shapes such as rectangular boxes, circles, or polygons of irregular shape. All features that fall inside the delineated shapes are then selected.

● Identify existing geographical regions. For example, selecting subdivisions from a city map causes all properties that fall inside the selected subdivisions to be selected.

The next illustration shows a typical example of selecting point features by identifying regions (polygons). In this case, three out of the nine regions in the map are identified graphically. The point features that fall inside the specified regions are then selected.

Point features within specified polygons can be identified and selected.

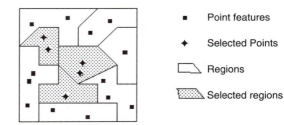

Alternatively, spatial features may be selected through queries or logical operations. Logical operations deal directly with the database and allow the user to identify and select features by a specific set of criteria. In most applications, features are identified and selected according to a combination of several conditions. For example, all parcels involved in transactions with a sale price greater than $200,000 over the past three years can be identified directly from the database and highlighted on the map. The selected features can be saved in a new coverage for further analysis.

In a typical application, a specific item in the database is employed to differentiate features that satisfy different sets of selection criteria. This item may record the combined conditions of sale price, transaction date, and other variables. New polygons are then delimited by dissolving the selected features according to the criteria.

Summary statistics can be derived either for an entire coverage or selected features. Basic summary statistics include the mean, variance, maximum, minimum, range, and frequency of every attribute specified by the user. These statistics typically provide the basis for spatial analysis. In addition, frequency distributions by attribute provide useful information about the data.

In most applications, frequency distributions are derived according to a given classification scheme. For instance, the frequency of objects within each class can be grouped, and the distribution pattern can be more clearly indicated by a frequency table. If the data are not pre-classified, then the entire range of the attribute will be used for depicting frequency distributions. Tables and charts are used to present frequency distributions.

In the next diagram, the three maps cover the same geographic area. Map A is classified by land use. According to the aggregate frequency, commercial land use accounts for 66 acres, and residential land use, 88 acres. The map of flood zone (B) shows that 34 acres are in the flood zone, and 120 acres in the non-flood area.

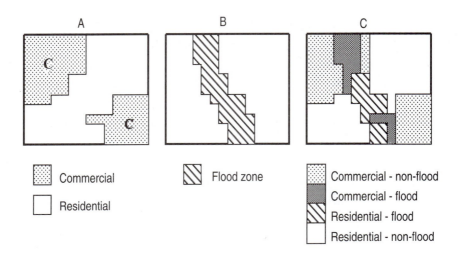

Maps showing land use, flood zone, and combinations.

Appearing below is the frequency table based on map C, the combination of land use and flood zone.

Land use and flood zone frequency table

Surface class	Frequency (acres)
Commercial, non-flood	48
Commercial, flood	18
Residential, non-flood	16
Residential, flood	12

With a GIS, the frequency table of any map can be generated based on one or more attributes. In the above example, frequency distributions were generated for land use, flood zone, and the two variables combined.

Feature Classification

Data classification is a common procedure in handling spatial information. Organizing data into classes according to specific attributes is particularly useful when you are working with numerous entities with a wide range of values for several attributes. Feature classification usually begins with the determination of the number of classes. In general, any number of classes can be specified, although psychological research into human capacity for handling information suggests that three to seven classes are the most effective.

At the left of the next illustration is a classification scheme for a polygon map in which the values of the study attribute range from 1 to 4. The simple map at the right is the result of defining two equal-interval classes.

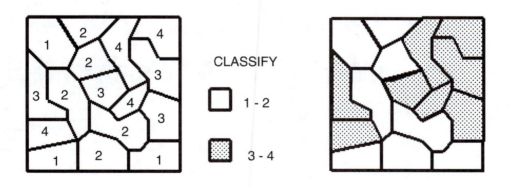

Classification of polygons produces a generalized map suitable for spatial analysis.

Once the number of classes is identified, several frequency distribution methods are available for classification. The following illustration shows the six common frequency distribution patterns—uniform, normal, bimodal, linearly skewed, non-linearly skewed, and multiple clusters.

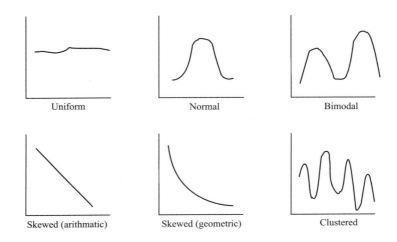

Six common frequency distribution patterns.

If the frequency distribution shows a relatively uniform pattern within the range, then a simple classification with a *constant interval* (equal-step series) can be applied. In this case, the maximum and minimum values of the data are evaluated, and the full range of the data is then divided into the determined number of classes. Formally, the class limits for the k-th class are specified as

$$\left[Z_{min} + (k-1)\frac{Z_{max}-Z_{min}}{n} ; Z_{min} + k\frac{Z_{max}-Z_{min}}{n} \right]$$

where Z_{max} and Z_{min} are the maximum and minimum of the attribute Z, respectively, and n is the number of classes.

For instance, if the maximum value of the attribute is 60 and the minimum value is 20, and if the data are to be classified into five classes, then the first class has limits of [20, 28], the second class has limits of [28, 36], and so forth. In this case, the interval for each class is a constant, while the frequency (number of objects) of each class may vary. If the distribution is uniform, then the variation in the frequency is minimal.

A second method is the *quantile* (equal frequency) classification. The principal objective is to maintain an equal or nearly equal frequency (number of objects) in each class. All objects are arranged in either ascending or descending order. The total number of objects is divided by the number of classes, and class limits are determined accordingly. A major disadvantage of this method is that class limits cannot be expressed mathematically because they are data dependent.

When a frequency distribution is similar to *normal* (bell-shaped), the classification can be based on the mean and standard deviation of the data. For instance, one class is specified by the mean plus one standard deviation, while a second class is specified by the mean minus one standard deviation. Two other classes are within the range between the mean plus one and mean plus two standard deviations. A third pair of classes are in the range between minus one and minus two standard deviations. In general, the number of classes pertaining to a normal distribution are limited to six, four, and two.

A *bimodal* distribution is treated as two separate groups. In this case, the data are first divided into two groups, and then each group is further classified by one of the above methods.

If the frequency distribution shows a *linear slope* pattern—the curve representing the histogram is skewed toward one end with a relatively constant slope—, the classification can be based on the arithmetic progression. In this case, a term can be defined to represent the interval of the k-th class as

$$T_k = b + (k-1)d$$

where b represents the first interval of the progression; d is the constant increment in the progression; k is greater than or equal to 1; and $T_0 = 0$. The limits for each class can then be defined as $[L_k, U_k]$ where a denotes the starting value of the first class, and

$$L_k = a + \sum T_{k-1}$$

$$U_k = a + \sum T_k$$

For example, if a = 0, b = 0, and d = 10, then the class limits for k = 1 are [0, 10]; for k = 2, [10, 30]; k = 3, [30, 60]; k = 4, [60, 100]; and so forth. If a = 0, b = 10 and d = 2, then the sequence is [10, 22], [22, 36], [36, 52], and so on. In the arithmetic progression series, the interval changes along the series to reflect the skewed pattern of the distribution.

If the slope of the skewed distribution is curvilinear instead of linear, then the classification can be based on a geometric progression. In this case, the class limits for the k-th class are

$$[ar^{(k-1)}, ar^{(k)}]$$

where a is the base value and r is a multiplier. For example, if a = 10 and r = 2, then the class limits for k = 1 are [10, 20]; k = 2, [20, 40]; k = 3, [40, 80], and so forth.

In reality, frequency distributions are usually more complicated than the patterns discussed thus far. The *multiple cluster* pattern represents a general case in which the curve is characterized by groups of varying width and multiple peaks of varying heights. Class limits cannot be expressed mathematically in this pattern. The optimal classification in such cases is one in which the class limits are determined by natural breaks in clusters.

To identify the bounds of the natural groups, the clustering method can be employed. There are several clustering methods available in commercial statistical packages. Among others, the hierarchical grouping method described below illustrates fundamental classification principles.

The *hierarchical clustering* method determines both the optimal number of classes and class limits. In the beginning, every object is treated as a one-member group. At each iteration, the individual object with the least difference from any existing group is identified and merged to that group. Thus, each iteration reduces the number of groups by one until all objects are grouped together in a single group. Three statistics, SS_w, SS_b, and the SS_w/SS_b ratio, are computed at every stage. SS_w is the sum of squared differences within each group, and SS_b is the sum of squared differences between groups. The grouping determines which objects are to be in the same class, and the number of natural clusters is determined from the curve of the SS_w/SS_b ratio.

The next illustration presents a typical pattern of the SS_w/SS_b ratio curve. The vertical axis represents the ratio, while the horizontal axis denotes the number of groups. At the right of the diagram where the number of groups equals the number of objects, SS_w equals 0 and the ratio is equal to 0. At the left, where the number of groups equals 1, SS_b equals 0 and the ratio becomes infinity. As a rule of thumb, the most suitable number of clusters can be identified from the "kink" or elbow on the curve. The rationale is that when the clustering starts from the far right of the curve, the increase in the ratio tends to be relatively slow until the curve hits the kink. After this particular point, the increase in the ratio becomes much greater than before.

A typical pattern of the SS_w / SS_b ratio.

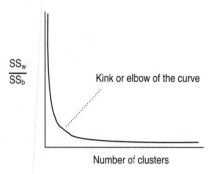

While statistical properties associated with each classification method can be analyzed, a critical question for spatial analysis is the efficacy of a particular classification for illustrating spatial patterns. The classification methods discussed above are all based on attribute data. In other words, all are statistical and aspatial. If the main purpose of classification is to study the spatial patterns and spatial processes of a phenomenon, then the result of classification must be evaluated with respect to the resultant spatial patterns.

In cases where map features are polygons, the *fragmentation index* can be used to evaluate classification results. Formally, the fragmentation index is defined as

$$\rho = \frac{m - 1}{n - 1}$$

where m denotes the number of contiguous map regions, and n denotes the number of original map units. For a polygon coverage, n is the number of polygons before classification and dissolving, and m is the number of polygons after classification and dissolving. The latter (m) is always less than or equal to n, and the value of the ρ index ranges between 0 and 1.

On one theoretical extreme, m = n, the classification creates a situation in which every polygon is in a class different from all adjacent polygons. In this case, the fragmentation index has a value of 1, implying the maximum level of fragmentation of the map features based on the classification. On the other extreme, all polygons are classified into the same class resulting in a single universe polygon. Because m = 1, the ρ index becomes 0, implying a completely consolidated pattern.

The following illustration demonstrates evaluation of a map pattern based on the ρ index. On the left is the original data for a 5x5 grid. The same data may be classified differently. Pattern A is the result of one classification scheme, and pattern B, the result of another. In both cases, the data are organized in three classes. Different class limits create different patterns. The ρ index for pattern A, a more fragmented pattern, is (8-1)/(25-1) = 0.29. For pattern B, the ρ index is significantly lower than that of pattern A, or (3-1)/(25-1) = 0.08.

*Different spatial patterns
for the same data
set can be derived from
different classification
schemes.*

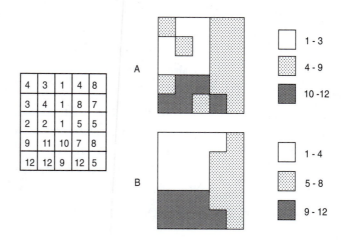

Summary

For convenience of discussion, data manipulation functions available in commercial GISs are classified into single layer operations and multiple layer operations. Most spatial analyses, however, involve both types of GIS operations. This chapter summarized the single layer functions most commonly employed in spatial analysis.

Single layer operations are essential for data preparation. In most cases, spatial data must be manipulated to generate appropriate data sets for coverage of a specific geographical area or to meet certain selection criteria. Preparation of spatial data involves three types of operational procedures—feature manipulation, feature identification and selection, and feature classification.

First, map features on a data layer can be spatially manipulated in various ways. A large landscape can be divided into subdivisions of smaller area while separate districts can be appended into large regions. Portions of a database can be removed, updated, or selected for further processing.

Second, map features can be effectively identified and selected from the database, either interactivlely or through logical queries. The identification and selection of spatial features enable the analyst to conduct comparative analyses based on any specified set of selection criteria.

Third, spatial objects must be classified before analysis or model building. Using a GIS, the frequency table of any selected set of spatial objects can be generated and the appropriate classification scheme can be employed.

Exercise

1. Digitize the following freeway and city boundary maps, either manually or using a GIS. When digitizing the freeway map, do not include the dashed line in the middle. This line represents a freeway under construction and is not ready for public use.

Freeways

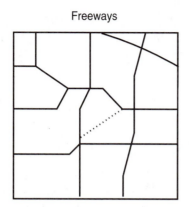

City boundaries

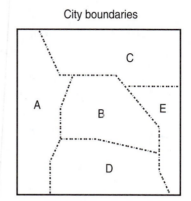

Hypothetical maps of freeways and city boundaries.

(a) Maintain the original coverage of city boundaries. Split the coverage by city using the SPLIT function described in this chapter.

(b) Clip the freeway map by the boundaries of City B to generate a new freeway map for this city only.

(c) Erase the freeways of City B from the original freeway map. Compare the outcome with step (b) results.

(d) Using the results of step (b), add the dashed line to the freeway map of City B. Then, update the original freeway map with the freeway map of City B.

(e) Create buffer zones extending from the updated freeways by a width equivalent to one-tenth of either side of the original freeway map. Define a new polygon coverage of a freeway access map by the buffer zones.

(f) Generate a frequency table to show the area of freeway access zones in each city.

2. Classify the polygons in the diagram below into three classes according to the specified criteria and compute the fragmentation index based on the following classification schemes: constant interval; equal frequency; 1 to 3, 4 to 9, 10 to 12; and 1 to 2, 3 to 10, 11 to 12.

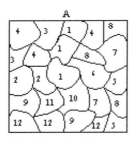

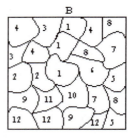

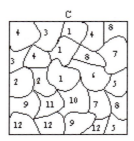

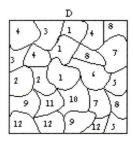

Multiple Layer Operations

This chapter focuses on analytical procedures that operate on multiple data layers. Multiple layer operations, also known as *vertical operations*, are based on the logical relationships among data layers. These operations provide the most fundamental tools for spatial analysis because they allow for the manipulation of data organized on separate layers and the examination of relationships among different features. With the use of these operations, you can separate data on the same layer into multiple layers in order to individually analyze separate yet related elements of a spatial phenomenon. In addition, multiple layers of geographical features that cover the same area can be combined (overlaid) to form a single layer for effective processing and model building.

As regards functionality in spatial analysis, multiple layer operations can be classified into the following three categories: overlay, proximity, and spatial correlation analyses. In general, overlay analysis involves the logical connection and manipulation of spatial data on separate layers. Proximity analysis deals with operational procedures that are based on distance measurement between features on different layers. Spatial correlation analysis is useful for revealing the relationships between features of different types. The three analysis types are discussed in subsequent sections.

Overlay Analysis

Overlay analysis manipulates spatial data organized in different layers to create combined spatial features according to logical conditions specified in Boolean algebra. The logical conditions are specified with operands (data elements) and operators (relationships among data elements). Common operators include *and, or, xor* (exclusive or), and *not.* Each operation is characterized by specific logical checks of decision criteria to determine if a condition is true or false. The following table shows the true/false conditions of the most common Boolean operations. In this table, A and B are two operands. One (1) implies a true condition and zero (0) implies false. Thus, if the A condition is true while the B condition is false, then the combined condition of A AND B is false, whereas the combined condition of A OR B is true.

True/false conditions of common Boolean operations

A	B	A AND B	A OR B	A NOT B	A XOR B
0	0	0	0	0	0
1	0	0	1	1	1
0	1	0	1	0	1
1	1	1	1	0	0

The most basic multilayer operations are union, intersection, and identity operations. All three operations merge spatial features on separate data layers to create new features from the original coverage. The main difference among these operations is in the way spatial features are selected for processing. For convenience, the command terminology of ESRI's ARC/INFO GIS is adopted in the following discussion. The command names employed for the same procedures in other GISs may be different.

UNION

The UNION procedure is equivalent to the Boolean OR operation in which two or more data layers are overlaid to produce a combined coverage. In essence, every feature on every layer is incorporated in the output coverage. The next diagram illustrates the UNION procedure. On the left is the input coverage with five polygons delineated by four rectangles. In the middle union coverage, three polygons are specified by two concentric circles. Combining these two coverages with the UNION procedure results in the output coverage on the right in which 14 polygons are defined. This operation requires that both the input coverage and the union coverage be defined by polygon features. Every polygon in the output coverage carries the attribute information of both input and union coverages. For instance, if the input coverage is a land use zoning map, and the union coverage shows school districts, then every polygon in the output coverage contains appropriate information for both land use and school districts.

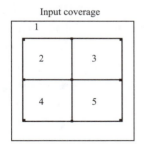
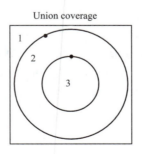
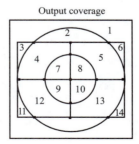

The input and union coverages are overlaid to produce the output coverage.

INTERSECT

The INTERSECT procedure is equivalent to the Boolean AND operation. When two coverages are processed, only the portion of the input coverage that falls inside the intersect coverage will remain in the output coverage. The additional information from the intersect coverage will be added to the output coverage as well. In brief, everything in the output coverage is present in both input and intersect coverages.

While the intersect coverage must be a polygon coverage, the input coverage can be a point, line, or polygon coverage. If the input coverage is a point coverage, the INTERSECT procedure will also result in an output coverage of point features. Likewise, a line coverage as input will result in a line coverage. For example, the INTERSECT procedure might be used to identify residential zoning areas that lie within 500 ft of major highways. In this case, the INTERSECT procedure applies to the land use zoning and highway buffer data layers. The results show areas that satisfy both the zoning and highway proximity conditions.

Input coverage

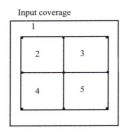

Intersect coverage

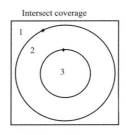

Output coverage

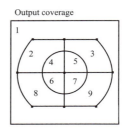

Intersecting multiple layers.

IDENTITY

When executing the IDENTITY procedure, every-thing located within the boundaries of the input coverage (including information in the input and identity coverages) is collected in the output coverage when merging multiple data layers. In other words, the outer boundary of the output coverage is identical to that of the input coverage. All information from the identity coverage within the outer boundary of the input coverage is added to produce the output coverage.

The identity procedure applies to coverages of points, lines, and polygons. However, if the input is a point coverage, the output coverage will contain only point features, although the identity coverage is always a polygon map. Likewise, a line coverage as input will result in a line coverage as output. In effect, the procedure requires that the "identity" of every feature in the input coverage be maintained. Assume, for instance, that you are working with two data layers—the street network for the entire state of California, and a land use map for Riverside (CA) county. Assume also that your analysis will focus only on the street network and land use of Riverside county. The identity procedure allows you to construct a combined coverage containing both the street network and land use for the county.

Input coverage   Identity coverage   Output coverage

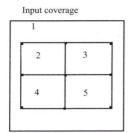 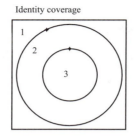 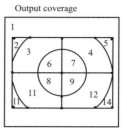

The identity of the input coverage is maintained through the IDENTITY operation by copying information from the identity coverage into the output coverage.

The following illustrations provide a typical example of overlay analysis. In order to examine the role of vegetation in the distribution of wildland fires in California's San Jacinto Ranger District, Richard A. Minnich (University of California, Riverside) interpreted aerial photographs and delineated three map layers of major vegetation types. The separate vegetation layers were then overlaid to derive a general vegetation classification appropriate for modeling the relationship between vegetation and wildfire distribution.

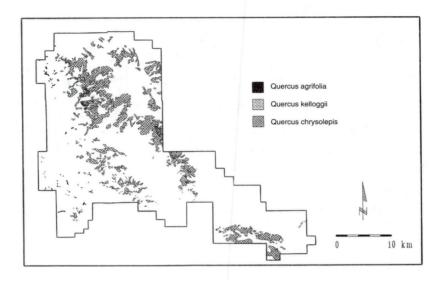

Forested areas dominated by oaks in San Jacinto Ranger District, California.

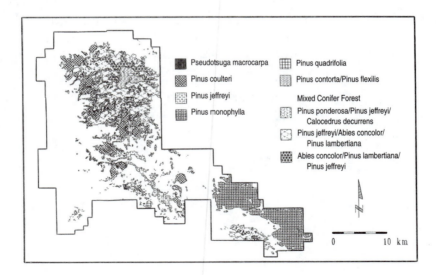

Distribution of conifer forests in San Jacinto Ranger District, California.

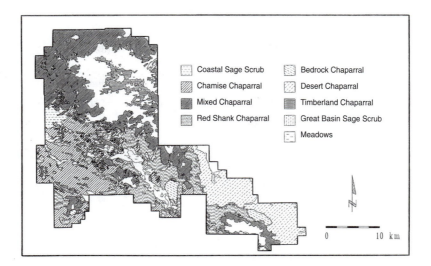

Distribution of understory species including chaparral and scrub in San Jacinto Ranger District, California.

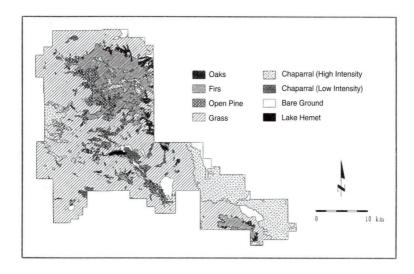

Overlay of the previous three vegetation layers producing a general vegetation classification appropriate for spatial analysis.

Frequency/Density

Spatial analysis often requires the calculation of the frequency count and/or density of features that are organized on one layer (*target layer*) with respect to geographic units delineated on another layer (*base layer*). Because the occurrence and frequency of the target features are evaluated in terms of a specific area, the base layer must be comprised of polygon features. The target layer can consist of point, line, or polygon features.

Assume that a county election official wants a frequency/density report of voters by precinct, to ensure that voters per precinct do not exceed the legislated maximum (e.g., 1,000). In this case, the base layer is a precinct boundary map (polygon coverage), and the target layer is a point coverage in which point features represent the address locations of registered voters. Frequency and density measures would permit the official to conduct a spatial analysis to study the impact of alternative redistricting plans. The next figure illustrates four existing precincts. Consequently, election officials are required to redraw precinct boundaries whenever voters surpass the maximum. With the use of a GIS, alternative redistricting plans can be delineated on the precinct boundary layer, while frequency and density statistics by precinct are computed automatically.

Electoral precincts can be redistricted based on frequency counts.

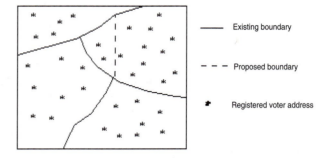

The previous example shows that point features on one layer can be counted with respect to the polygon features delineated on another layer. In brief, frequency measures are derived for geographic units. Point feature density can be calculated by overlaying the point and polygon coverages and dividing the point feature frequency by area units.

Line features are one-dimensional, and thus, a frequency measure based on the number of line segments is not appropriate. In general, length is more suitable for evaluating the properties of line features. The total length of all line features on the target layer that fall inside a geographic unit on the base layer can be computed as a frequency measure. The total length divided by the area of the corresponding polygon is thus a valid density measure of line features. For instance, a nation's transportation development is commonly measured by either the total length of railroads and/or freeways, or the density of such features in miles per square mile.

Because polygon features are two-dimensional, frequency measures must be based on area. The next illustration shows an example of area frequency computations in a GIS. The diagram at the left represents a land use zoning map, and the diagram in the center, a management district map. When these two layers are overlaid to form the map on the right, land use zones can be computed by district.

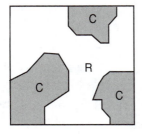 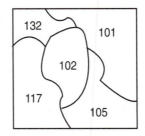 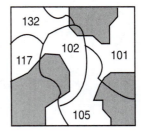

Polygon feature area frequencies can be derived through map overlays.

In the above example, there are two types of land use zones, residential (R) and commercial (C). The five management districts are labeled 101, 102, 105, 117, and 132. The map overlay of the two layers generates a table showing the total area of residential and commercial zones in each district.

Area frequency of land use zones by management district

District	Land use	Area
101	Residential	408
	Commercial	333
102	Residential	272
	Commercial	156
105	Residential	353
	Commercial	209
117	Residential	287
	Commercial	406
132	Residential	272
	Commercial	12

Proximity Analysis

Distance between different types of spatial features is the primary element of proximity analysis. For instance, the distance between a land parcel and the nearest highway is considered a significant measure of transportation accessibility, and thus could be used as an indicator of land value. Proximity analysis results alone do not usually serve as the sole purpose of a spatial analysis. In most cases, the proximity analysis is conducted to generate information about spatial properties used in other applications. An example is determining whether residential property value is related to distance from commercial establishments. Before this question can be answered, the analyst must calculate the distance between every residential parcel and the nearest commercial establishment(s) in the study area. Parcel data are typically organized in one layer, and commercial establishments are on another. Thus, proximity analysis is used to calculate distance from every parcel on a parcel map layer to the corresponding nearest commercial establishment on a land use layer.

Proximity is generally calculated in different ways according to feature type. The proximity between polygons can be evaluated either by *interseparation distance* (i.e., the shortest separation between polygon perimeters) or the *distance between centroid locations*. Interseparation distance is measured more effectively through grid analysis, which is discussed in Chapter 10. Measuring proximity between centroid locations is similar to the point-to-point distance measure discussed in this chapter. Proximity measurement for line features is accomplished through network analysis, the main topic of Chapter 7. In this chapter, proximity measurement refers to point features only.

NEAR

The most fundamental function of proximity analysis in GIS is the NEAR function which identifies the nearest point or line feature on one layer from a point feature on another layer, and computes the distance accordingly. The next illustration demonstrates typical operations of this function.

The NEAR function finds the distance between a point on one coverage and its nearest line segment on another coverage.

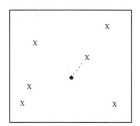 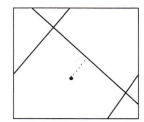

The diagram on the left side shows the NEAR function applied to point features only. In this example, the dot in the center represents the point on the base layer, and the Xs represent the points on the target layer. The NEAR function finds the point on the target layer closest to the dot, and computes the distance between the dot and the nearest X.

For instance, the dot may represent a building on fire where buildings are shown as point features on the source layer, while the Xs may represent fire stations. The NEAR procedure would identify the fire station closest to the building on fire. At the right of the diagram, proximity is evaluated between the point (the dot in the center) on one layer and line features on the other layer. For instance, the dot may represent the site of a new building where connection to water service is required, while line features represent water supply pipelines. The shortest connection between the building and the pipe can be identified using the NEAR function.

POINTDISTANCE

The POINTDISTANCE procedure identifies all point features on one layer that are within a specified distance range from each point on another layer, and computes the distance between those points and each of the identified points. This procedure may also be applied to point features in the same data layer. The following diagram illustrates an example of this function. The dot in the center represents a feature on the base layer, while the Xs represent point features on another layer (the target layer). In this operation, a search radius is specified and the POINTDISTANCE function identifies all points on the target layer within the specified radius from each point on the base layer. The straight line distance between each point on the base layer and every identified point on the target layer is computed.

POINTDISTANCE identifies all points on a coverage that are within a specified distance from a point on another coverage.

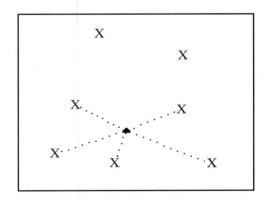

The previous examples illustrate that the distance between the points on one layer and points on another can be evaluated. The distance measure based on point features may take one of the following forms.

- *One-to-one.* Provides the distance between a point on the base layer to the nearest point on the target layer.

- *One-to-many.* Provides the distance between a point on the base layer to every point on the target layer.

- *Many-to-many.* Evaluates the distance from every point on the base layer to its corresponding nearest point on the target layer.

- *Many-to-all.* Evaluates the distance from every point on the base layer to all points on the target layer.

Likewise, the distance measure between point features on the base layer and line features on a target layer may take the form of one-to-one, one-to-many, many-to-many, or many-to-all.

Spatial Correlation Analysis

The primary objective of spatial correlation analysis is to reveal relationships between different types of spatial features. Spatial correlation analysis determines whether the distribution of one type of feature, organized in a particular data layer is related to the distribution of features organized in another data layer.

For instance, an analyst might wish to determine whether soil and vegetation distributions are related.

The relationship between multiple types of spatial features provides useful information for the explanation of the distribution of any feature type. If vegetation and soils are highly correlated, then the vegetation map may be used as an indicator of the underlying soils, and the soil map may be used to delineate potential areas of certain vegetation types.

Furthermore, understanding the correlation between different features is important for spatial modeling. For instance, if two layers are highly correlated, then the information on these layers may be redundant, and using both features to explain the same phenomenon is unnecessary. Therefore, if vegetation and soils are highly correlated, then a model of the distribution of wildfires need not incorporate both variables.

Spatial correlation analysis is useful for a wide variety of applications. In the realm of business applications, assume that the marketing department of XYZ Foods knows that a particular food product is preferred by Asian Americans. The company can use census data to delineate geographical areas of high concentration of this ethnic group in order to define the most effective spatial advertisement strategy.

Spatial correlation analysis between different feature types is executed in different ways, depending on the measurement scale of study variables. When the variables are nominal scale, spatial correlation can be analyzed using the contingency table and the χ^2 test of goodness-of-fit. If the variables are interval or ratio scale, then the correlation coefficient and regression models provide a more appropriate method for correlation analysis.

Contingency Table and χ^2 Test of Goodness-of-Fit

The frequency measurements described in Chapters 3 and 4 are useful for showing the spatial pattern in the distribution of a phenomenon. In addition, the measurements can be used for testing hypotheses about the relationship between different features. With correlation analysis, you can investigate whether a frequency tends to be higher in certain areas.

For instance, a criminologist might be interested in whether crime incidence is related to the ethnic distribution of a city's population. Wildlife biologists may need to know if an endangered species is more likely to be found in specific areas of vegetation cover rather than in others. For questions like these, the study variables could be categorical (nominal scale), such as vegetation type or ethnicity. In this case, GIS operations can determine whether two layers of polygon features are correlated.

The following table presents a hypothetical example of the relationship between vegetation cover and the distribution of animal species. The vegetation cover consists of three distinct types labeled V1, V2, and V3. The area is inhabited by three animal species, labeled A1, A2, and A3. The vegetation coverage contains polygons that represent areas of relatively homogenous vegetation types. The habitat coverage is a point coverage in which each point denotes the site where the specific species is observed. Overlay of these two layers produces a frequency distribution table called a *contingency table.* At the left of the table below, a 3X3 matrix shows the observed distribution of the animals with respect to the three vegetation types.

Contingency table

		Observed distribution						Expected distribution			
		Vegetation type						Vegetation type			
		V1	V2	V3	Σ			V1	V2	V3	Σ
	A1	75	10	15	100		A1	29.3	37.3	33.3	100
Animal species	A2	5	110	20	135	Animal species	A2	39.6	50.4	45.0	135
	A3	30	20	90	140		A3	41.1	52.3	46.7	140
	Σ	110	140	125	375		Σ	110	140	125	375

Columns in the contingency table represent the observation frequency of the three species in three vegegation type polygons. Rows represent the observation distribution of a specific species among the three vegetation types. For instance, the column1-row1 cell indicates that 75 animals of species A1 were observed in vegetation V1. The principal research question here is whether the distribution of animal species is related to vegetation type. The contingency table is constructed to detect relationships between these two layers.

For example, if species A1 is closely associated with vegetation V1, a disproportionally high frequency for the A1-V1 cell is expected, compared to the A1-V2 and A1-V3 cells. If, on the contrary, there is no relationship between A1 and vegetation type, then a rather random distribution in the contingency table is expected. In order to determine if a relationship exists between species and vegetation type, another frequency distribution table is necessary, based exclusively on randomness. In this table, it is assumed that the expected frequency of each cell in the table is determined by the frequency of every row and column. This frequency table contains the expected random distribution of the species if there is no relationship between species and vegetation type.

To build the expected distribution table, the assigned value of each cell is computed by the product of row sum and column sum. The product is then divided by the total frequency. For example, the expected frequency of animal species A1 in vegetation type V1 is equal to $(100 \times 110)/375 = 29.3$. When every cell is computed according to its corresponding row and column, the expected frequency table is complete. (See right side of previous table.) The next table lists the computed quantities needed for the χ^2 test of goodness-of-fit.

Statistics derived for χ^2 test of goodness-of-fit

I	O_i	E_i	$O_i - E_i$	$[O_i - E_i]^2$	$[O_i - E_i]^2/E_i$
1	75	29.3	45.7	2088.49	71.28
2	10	37.3	-27.3	745.29	19.98
3	15	33.3	-18.3	334.89	10.06
4	5	39.6	-34.6	1197.19	30.23
5	110	50.4	59.6	3552.16	70.48
6	20	45.0	-25.0	625.00	13.89
7	30	41.1	-11.1	123.21	2.99
8	20	52.3	-32.3	1043.29	19.95
9	90	46.7	43.3	1874.89	40.15
				Sum	279.01

The previous table shows the procedure for computing the statistic $\Sigma \ (O_i - E_i)^2 \ /E_i$, which has a χ^2 distribution with (u-1)(v-1) degrees of freedom; u stands for the number of rows in the contingency table and v is the number of columns. The purpose is to test the hypothesis that there is no significant relationship between the two data layers and that the frequency distribution is the result of a random process. If the derived statistic exceeds the 95% confidence interval of the χ^2 distribution, the null hypothesis that animal species are distributed randomly in the region should be rejected. In this case, the alternative hypothesis is that animal species are distributed unequally and have a tendency to be found more frequently in certain vegetation types than others.

In the example, u = 3 and v = 3; therefore, the statistic has (3-1)(3-1) = 4 degrees of freedom. In the

χ^2 distribution, P(χ^2 < = 0.99) for 4 degrees of freedom = 13.28. Because the computed statistic (279.01) far exceeds the specified critical value at 99% confidence level, the null hypothesis—the distribution of animal species over vegetation types is random—is rejected. In other words, this example suggests that the distribution of animals is related to the distribution of vegetation. In most applications, a .95 confidence level is sufficient for rejecting the null hypothesis of random distribution.

Correlation Coefficient

The preceding section demonstrates how the correlation between two data layers can be analyzed if the variables involved in both layers are measured at a nominal scale. When the study variables are measured at an interval or ratio scale, the spatial correlation between different feature types is more appropriately tested by Pearson's correlation coefficient or regression analysis. While the Pearson correlation coefficient is useful for showing the correlation between data layers, in spatial analysis, regression models are generally more useful because the parameters estimated in the regression are needed for calibration and application of spatial models.

In other words, the correlation coefficient only indicates the extent to which the value of another variable is related to the value of the other variable, whereas regression analysis provides additional information about how the relationship is expressed mathematically. For instance, although a high, positive correlation coefficient indicates that variables x and y are significantly correlated, it does not indicate how a unit change in one variable affects the other. A simple regression would yield a quantitatively expressed function, such as y = 2x. In this example, the estimated parameter provides some useful information that a one-unit increase of variable x implies an increase of 2 units in variable y.

The next table lists the attribute table of a hypothetical parcel map coverage in which eight parcels are registered. The first column lists parcel IDs and the second column denotes property value. The property value of each parcel is copied from the sale price recorded on the original parcel map. The third column contains the distance from the centroid location of each parcel to the nearest metro station. Metro station locations are recorded on another data layer. The distance to nearest metro station for each parcel was obtained using the basic multiple layer operations described in preceding sections. Because both variables are measured at a ratio scale, the relationship between property value and distance to the nearest metro station can be analyzed using the correlation coefficient.

Parcel ID	Property value (000)	Distance to metro station (miles)
101	31	0.75
102	31	1.50
103	29	1.75
104	27	2.50

Parcel ID	Property value (000)	Distance to metro station (miles)
105	27	3.25
106	26	3.75
107	25	4.50
108	23	5.00

The relationship between property value and distance to metro station can be examined from the next illustration, which is derived from the preceding table. In the diagram, the vertical axis represents sale price and the horizontal axis represents distance to the nearest metro station. Every dot in the diagram represents a parcel plotted according to its corresponding values. The diagram shows an inverse relationship between these two variables—greater distance implies 'lower property value. To confirm this relationship and make it precise, Pearson's correlation coefficient can be used.

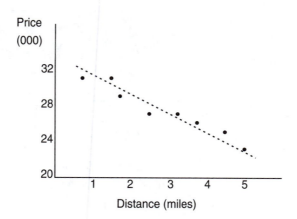

Hypothetical example relating property value and distance to metro stations.

The correlation coefficient (r) between two variables, x and y, is defined as follows:

$$r = \frac{\Sigma_i(x_i - \bar{x})(y_i - \bar{y})}{nS_xS_y}$$

where n is the number of geographic units (n = 8 in previous example), and S_x and S_y denote the standard deviation of variables x and y, respectively.

The correlation coefficient (r) indicates the degree and direction of the relationship between two variables. The values of r range between -1 and 1. A positive r value indicates that there is a positive (direct) correlation between the two variables, that is, a larger value of one variable implies a larger value of the other. At the extreme, r = 1, the two variables are perfectly correlated and their distribution patterns must be identical. A negative r value implies a negative (indirect) correlation; the variables are inversely related to each other. A higher value of one variable implies a lower value of the other. If the r value is not significantly different from 0, there is no correlation between the variables and the distributions of the two variables are considered independent of each other.

In the example, the negative correlation between property value and distance from the nearest metro station is obvious. The computed r equals -0.9695, indicating a highly significant negative correlation between the two variables.

Simple Linear Regression

In the previous illustration, the relationship between property value and distance to the nearest metro station can be represented by a linear function, such that

$$P = a + bD$$

where P represents property value and D denotes distance from the nearest metro station; b is a parameter associated with the correlation between P and D; and a is a constant indicating the vertical intercept of the line expressed above. In the example, a denotes the height on the y axis (property value) where the line intersects, and b indicates the slope of the line.

Expressed in general terms, the simple regression model for analyzing the linear relationship between two variables follows:

$$y_i = a + b_{xi} + e_i$$

where a is the vertical intercept, b represents the slope of the line, and e is a randomly distributed error term.

Let y_i represent the observed value of the i-th record of the variable y, and let y_i' denote the estimated value of the same record. The best linear, unbiased estimates of parameters a and b can be obtained by minimizing the sum of differences between the observed points and the representative line in the previous figure, such that

$$\text{Minimize } \Sigma_i e_i^2 = \Sigma_i (y_i - y_i')^2$$

The estimates can be derived as follows:

$$b = \frac{\Sigma_i (x_i - \bar{x})(y_i - \bar{y})}{\Sigma_i (x_i - \bar{x})^2}$$

and

$$a = \bar{y} - b\bar{x}$$

The regression model is usually evaluated using the *coefficient of determination* (r^2), such that

$$r^2 = \frac{\Sigma_i (y_i' - \bar{y})^2}{\Sigma_i (y_i - \bar{y})^2}$$

Accordingly, the example of the relationship between property value and distance from the nearest metro station is expressed in a linear regression as

$$P = 32.60 - 1.82D$$

with an r^2 of 0.94, which suggests a highly significant correlation.

This chapter described simple linear regression between variables derived from two separate data layers. In more complex analyses, multiple variables are involved and the regression incorporates multiple variables. Multiple regression analysis is discussed in Chapter 8.

Summary

A fundamental requirement of GIS is the capability to process spatial data organized in separate, yet logically connected layers. The distribution of a geographic phenomenon tends to be affected by several factors that are interrelated in complex ways. The main purpose of spatial analysis is to sort out such factors and identify significant factors in the distribution of the phenomenon under consideration. Because spatial data are most effectively organized in separate layers, multiple layer operations become a required part of almost every spatial analysis.

Multiple layer operations are classified into three main analysis categories: overlay, proximity, and correlation. Procedures in overlay analysis are derived from Boolean algebra. GIS overlay functions ensure that spatial data organized in separate layers can be combined in any appropriate form needed for the analysis. The primary objective of overlay analysis is to build the logical connections among separate data layers in order to establish the relationships between features that represent different factors.

Proximity analysis deals with the most fundamental topic in geographical analysis, the role of distance in spatial relationships. Correlation analysis reveals relationships between a spatial phenomenon and distributions of significant variables.

In general, most GIS-based spatial analyses involve all three types of multiple layer operations.

Exercise

1. Digitize the distributions of soils and vegetation depicted below. In these maps, soil polygons are labeled by soil type while vegetation polygons are classified into grass (G), conifers (C), and oaks (O). Overlay the coverages with the "union" function. Generate a frequency table based on the overlay, and then construct a contingency table.

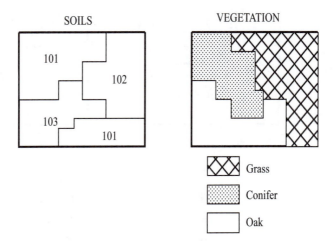

SOILS

VEGETATION

Grass
Conifer
Oak

2. Digitize the land ownership and flood zone coverages depicted below. Overlay the two coverages with the "identity" function based on land ownership polygons. Given that the annual rate of flood insurance is differentiated by flood zone—$30/acre for zone 1, $40/acre for zone 2, $50/acre for zone 3, and $60/acre for zone 4—, generate the land ownership frequency table and calculate the annual premium for each landowner.

Land ownership

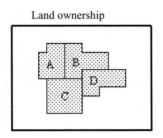

Flood zones

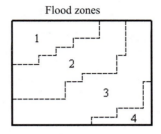

▨ 1 acre

3. The following point coverage shows the locations of commercial properties involved in recent transactions. The line coverage represents freeways in the same geographic area. Digitize both maps either manually or using a GIS. Apply the "near" function to identify the distance from every property to its nearest freeway. The sale prices of the properties are shown in the table below. Accordingly, build the regression model Price = a + b(Distance); estimate the parameters, a and b; and find the r^2 of this regression.

Property Number	Price (000)
1	120
2	180
3	250
4	250
5	100
6	120
7	300
8	220
9	270
10	180
11	280
12	160
13	180
14	130
15	180

4. According to the frequency table below, determine whether the correlation between vegetation and soil is significant using the χ^2 test of goodness-of-fit (χ^2 = 6.635 at the 99% confidence level for d.f. = 1).

Soil	Vegetation	Frequency
A	101	115
A	102	22
B	101	23
B	102	15

5. A contingency table was constructed from the frequency distribution of a polygon coverage where polygons are coded with two attributes, land use and principal ethnicity. The goodness-of-fit test was conducted to determine whether the two attributes are correlated. Because there are three ethnicity types and four land use classes, the contingency table consists of three rows and four columns. Consequently, there are (3-1)(4-1) = 6 degrees of freedom.

The χ^2 statistic at 99% level for 6 degrees of freedom is 16.81. The $\Sigma [O_j - E_j]^2/E_j$ quantity computed from the contingency table is equal to 12.66 (where O_j represents observed frequency for the j-th cell in the table and E_j represents the expected frequency for the corresponding cell).

According to the computed quantity, are land use and ethnicity correlated? Explain your conclusion.

Point Pattern Analysis

A point pattern is defined as the spatial pattern of the distribution of a set of point features. In point pattern analysis, spatial properties of the entire body of points are studied rather than the individual entities.

Because points are zero-dimensional features, the only valid measures of point distributions are the number of occurrences in the pattern and respective geographic locations. Area is not a valid measure, even though in most cases point features on a map actually occupy space. When geographic entities are represented as points on a map, all points are considered to be of the same quality. In other words, an analysis of point patterns is focused on the spatial pattern in the distribution of the point features, rather than variation in quality of the point features. This chapter first introduces common descriptive statistics of point features, and then presents selected methods for evaluating spatial patterns in the distribution of point features.

Descriptive Statistics of Point Features

The distribution of point features can be described by frequency, density, geometric center, spatial dispersion, and spatial arrangement. With the exception of spatial arrangement, evaluation of the spatial properties of point features can be grounded on basic descriptive statistics. This section is focused on descriptive statistics of point patterns. Issues pertaining to spatial arrangement will be discussed in the next section.

Frequency is the number of point features occurring on a map. Frequency is always the first measurement of a point distribution whenever two distributions of point features are compared, or when the same distribution of a point pattern is evaluated at different times in order to study the pattern's developmental process. However, the comparison of two distribution frequencies may be misleading if area is not considered. When two point patterns that differ in area are compared, it is prudent to evaluate the patterns by density, which is the ratio of frequency to area.

The next illustration shows four hypothetical point patterns with varying geometric centers and dispersions. These patterns are labeled A, B, C, and D. The following table contains the x and y Cartesian coordinates of all point features in each distribution.

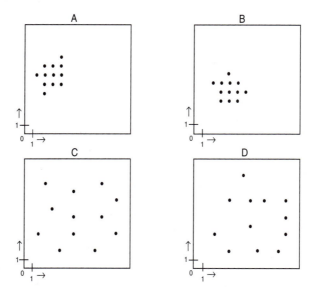

Four hypothetical point patterns with varying geometric centers and dispersions.

Cartesian coordinates of point features

A	B	C	D
(2, 7)	(3, 6)	(2, 4)	(3, 4)
(3, 5)	(4, 4)	(3, 10)	(5, 2)
(3, 6)	(4, 5)	(4, 7)	(5, 8)
(3, 7)	(4, 6)	(5, 2)	(7, 11)
(3, 8)	(5, 4)	(7, 4)	(8, 5)
(4, 6)	(5, 5)	(7, 6)	(8, 8)
(4, 7)	(5, 6)	(7, 9)	(9, 2)
(4, 8)	(5, 7)	(10, 2)	(10, 8)
(5, 6)	(6, 4)	(11, 6)	(12, 2)
(5, 7)	(6, 5)	(11, 10)	(13, 4)
(5, 8)	(6, 6)	(13, 4)	(13, 6)
(5, 9)	(7, 5)	(13, 8)	(13, 8)

Frequency is identical for all four distributions. In each case there are 12 point features. If the area of each distribution is equal to 180 units, then the density is 12/180 for every distribution.

The geographical properties of a point pattern are characterized by geometric center and dispersion. Important factors include how the points are distributed over the map area and whether the points are near the center of the map or clustered near a corner. Because every point feature is of equal weight, the geometric center of a distribution can be measured by the means of x and y coordinates, while spatial dispersion can be measured by the standard deviation of each mean. In other words, the geometric center of a point distribution is represented by a point location specified by the mean of x and y coordinates. In the following illustration, geometric centers of the four distributions are located at (3.83, 7), (5, 5.25), (7.75, 6), and (8.83, 5.67), respectively.

Geometric centers of the four point distributions in previous illustration.

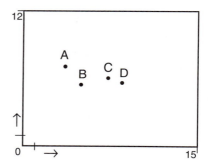

The meaning of geometric center for a point distribution is dependent on its dispersion. For instance, geometric centers for distributions A and B adequately represent central tendency because both distributions are concentrated around respective centers. In contrast, distributions C and D illustrate a widespread pattern, and thus their geometric centers are not reliable indicators of these distributions.

Because maps are two-dimensional, the dispersion of a point distribution is related to two quantities, the standard deviations of the x and y coordinates. The former indicates the dispersion of the points along the x axis, whereas the latter indicates the dispersion along the y axis. In the first illustration, the dispersion of pattern B, a concentrated distribution, is indicated by the low standard deviation values of 1.08 and 0.92 for x and y, respectively. For the scattered distribution of pattern C, the dispersion is indicated by larger standard deviation values of 3.65 and 2.74 for x and y, respectively. Clearly, the larger the standard deviation value, the greater the distribution's dispersion.

Furthermore, geometric center is not a reliable indicator of central tendency when either standard deviation has a large value. Because the dispersion along the x axis is independent from the dispersion along the y axis, it is possible for a distribution to show a large standard deviation on one axis and a small standard deviation on the other. In this case, the distribution would illustrate a linear pattern along one axis, and the point pattern may form an elliptic shape. A more advanced analysis of the point pattern requires either the rotation of the coordinate system or the computation of the correlation coefficient between x and y in order to more accurately represent the pattern.

The next illustration shows four point patterns that differ in dispersion characteristics. Pattern A illustrates a large standard deviation along the x axis and a lower level of dispersion along the y axis. Consequently, the distribution is elongated horizontally. When the dispersion is small along the x axis and large along the y axis, the distribution becomes elongated vertically, as shown in pattern B. When both x coordinates and y coordinates show a similar level of variance, the correlation between x coordinates and y coordinates can be examined to determine the pattern. For instance, if the correlation coefficient is positive and significant, then the point features are distributed in an elongated, sloping pattern similar to pattern C. If the correlation coefficient is insignificant while both standard deviations are similar, the distribution is a random pattern similar to pattern D.

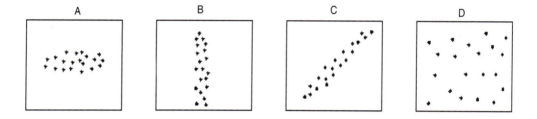

Four point patterns of different variance characteristics.

Spatial Arrangement

The spatial arrangement of point features is an important characteristic of a spatial pattern, because the location of point features and the relationships among them have a significant effect on the underlying process generating a distribution. The three basic types of point patterns are listed below.

- *Clustered.* Point features are concentrated on one or a few relatively small areas and form groups.

- *Scattered or uniform.* Characterized by a regularly spaced distribution with a relatively large inter-point distance.

- *Random.* Neither the clustered nor the scattered pattern prevails.

The following illustration displays three typical patterns of spatial arrangement.

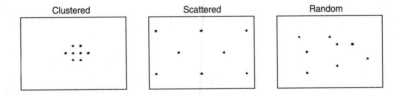

Three typical point distribution patterns.

Spatial arrangement in empirical point patterns can be more complicated than the three general classes. For instance, a distribution may illustrate a scattered system of point pairs, that is, two points are adjacent while the space between pairs (interspace) is relatively large. Another example is small groups distributed in a scattered pattern. Certain bird species show a pattern in which each mated couple occupies a territory during the mating season, while the distribution of the pairs may be scattered. Families with youngsters then form a small group, although the between-group distance remains large.

Spatial arrangement can be measured in different ways. The most commonly employed method is the nearest neighbor index.

Nearest Neighbor Index

The measurement of spatial arrangement can be based on the distance between adjacent point features. The nearest neighbor index measures the degree of spatial dispersion in the distribution based on the minimum of inter-feature distances. The rationale is that, in general, the average distance between points in a clustered pattern is shorter than in a scattered pattern. In addition, a random pattern is associated with an average inter-space distance larger than a clustered pattern and smaller than a scattered pattern.

To evaluate the nearest neighbor index, the nearest neighbor for each point feature must be found, and the inter-feature distance computed. If d_i denotes the distance from point i to its nearest neighbor, then

$$Ad = (\Sigma_i d_i)/n$$

is the average nearest neighbor distance of the point pattern, where n is the total number of points in the map area.

The coordinates of the eight points of pattern A in the preceding illustration appear in the following table.

A	B	C
(6,5)	(1,2)	(3,7)
(7,6)	(1,8)	(4,2)
(7,5)	(4,5)	(4,5)
(7,4)	(7,2)	(7,7)
(8,6)	(7,8)	(8,3)
(8,5)	(10,5)	(8,6)
(8,4)	(13,2)	(10,6)
(9,5)	(13,8)	(12,4)

According to the above coordinates, the nearest neighbor for every point in a pattern can be identified, the nearest neighbor distance computed, and then the average of all nearest neighbor distances computed. In this example, Ad is equal to 1 for pattern A, 4.24 for pattern B, and 2.26 for pattern C. Clearly, a smaller Ad number indicates a closer distance between points.

Assuming that the point pattern is random, the expected value of the average nearest distance, Ed, is expressed as

$$Ed = \frac{1}{2}\sqrt{\frac{A}{n}}$$

where A denotes the map area.

The expected distance is derived from the ratio A/n; that is, this quantity represents the average area each point would occupy in an evenly distributed system. The square root of this quantity converts the area measure into the interseparation distance between a pair of adjacent points. However, this interseparation distance represents only the expected interseparation in a regular pattern.

In a clustered pattern, the two adjacent points may overlap; thus, the distance becomes zero. The expected average in a random pattern, Ed, is therefore one half of the average interseparation distance. In the above example, all three patterns in the preceding illustration have identical frequencies (n = 8) and areas (A = 140). Consequently, Ed is equal to 2.09 for all three patterns.

The nearest neighbor index (NNI) is defined as the ratio of Ad to Ed as follows:

NNI = Ad / Ed

The values of NNI range between two theoretical extremes, 0 and 2.1491. When all points in a pattern fall at the same location, the pattern represents the theoretical extreme of spatial concentration. In this case, Ad = 0 and NNI = 0.

In a scattered pattern, the interseparation distance increases. When Ad approximates its upper bound, NNI approaches the theoretical maximum of 2.1491. When NNI is closer to 1, the pattern is considered random because Ad is about equal to Ed, which represents the average interseparation distance in a random distribution. In general, a smaller NNI value indicates a clustered pattern, whereas a larger NNI represents a scattered pattern. In the preceding illustration, NNI for pattern A is 0.41; for pattern B, 2.03; and, for pattern C, 1.08.

To test the statistical significance of a computed NNI, the standard normal deviate, z, can be derived as follows:

$$z = (Ad - Ed)/\sigma_{Ad}$$

where σ_{Ad} is the standard deviation of Ad computed by

$$\sigma_{Ad} = \sqrt{\frac{0.0683 A}{n^2}}$$

The computed z value can be compared with the normal distribution value of 1.96 for $\alpha = .05$, to test the hypothesis of whether the spatial pattern is random. If the computed z is greater than the critical value of 1.96, the point pattern is significantly different from that of a random process.

The NNI has been used extensively for evaluating the spatial dispersion of point patterns, and the computation of this index is relatively straightforward and simple. Nevertheless, this index is not sensitive to complex patterns. For instance, if distribution of an animal species is characterized by families with established territory, the interseparation distance is small within each family, yet the distance between families is large. In this instance, the computed NNI is not different from a pattern dominated by a single cluster.

When working with complex patterns, the NNI must be extended to the distance measured from the second nearest neighbor, the third nearest neighbor, and so forth. In doing so, the computation becomes cumbersome and the advantage of a single indicator for a spatial pattern no longer holds.

Quadrat Analysis and Poisson Process

Quadrat analysis of point patterns requires overlaying quadrats onto a map of point features in order to examine the distribution based on frequency of occurrence rather than interseparation distance. In this case, the evaluation of spatial patterns is determined by the frequency of points within each quadrat. Several approaches are available for conducting quadrat analysis. The quadrats can be different sizes or shapes, and may be placed on the map either randomly or regularly to form a grid.

Once quadrats are placed on the map, the frequency of points within the boundaries of each quadrat is counted. All quadrats are then classified according to the observed frequency within quadrats. In order for the statistic derived from a quadrat analysis to be valid, each class must contain at least five point features. In a point distribution, if another quadrat contains just one point, another quadrat contains two points, and three quadrats contain three points, then the quadrats containing one to three points must be combined to form a class to ensure that there are at least five occurrences in the class.

For each class, the observed frequency is obtained and defined as O_i, where i stands for the class i. The probability of occurrence for each quadrat, given the Poisson distribution, is then computed as

$$P(x) = \frac{\lambda^x e^{-\lambda}}{x!}$$

where x is the frequency in a quadrat, λ is equal to the expected frequency of the corresponding quadrat, and e is the base of natural logarithm, 2.718282.

The expected frequency value can be computed for each class i as E_i. The statistic

$$\chi^2 = \Sigma_i (O_i - E_i)^2 / E_i$$

has a χ^2 distribution with degrees of freedom equal to the number of classes minus two. If the computed quantity of χ^2 has a value less than the statistic obtained from the table, then the hypothesis that the pattern is random is not rejected.

The next illustration shows three distinct point patterns for a quadrat analysis. In pattern A the point features are regularly spaced and form a scattered distribution. Pattern B illustrates a random case. In pattern C the point features are clustered to form a single major group. Each distribution is overlaid onto a 6x6 grid resulting in a configuration of 36 grid cells.

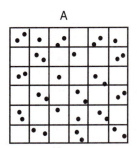
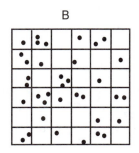
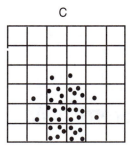

Three point distributions of distinct spatial patterns.

The quadrat analysis is based on the null hypothesis that point features are randomly distributed in the Poisson process. The following table shows the computation of required statistics for pattern A in the preceding illustration.

Quadrat analysis of point pattern A

Density (x)	Observed O_i	Probability P(x)	Expected E_i	$(O_i - E_i)^2 / E_i$
0	15	.3679	13.24	0.23
1	6	.3679	13.24	3.96
>=2	15	.2642	9.52	3.15

According to the previous table, the computed χ^2 quantity for pattern A is 0.23 + 3.96 + 3.15 = 7.34. Because the calculated quantity is beyond the critical value of the χ^2 at the .99 confidence level, the null hypothesis of a random distribution is rejected. Consequently, pattern A is significantly different from a random pattern. Likewise, the χ^2 quantities for patterns

B and C can be computed in the same way to determine if either pattern is the result of a random process.

Spatial Autocorrelation

Every method for evaluating spatial arrangement has limitations. Therefore, evaluating spatial arrangement based on one method alone may not be appropriate. The nearest neighbor index is derived from the nearest interseparation distance, but does not consider overall spatial distribution. For instance, this index would not differentiate between the two distinct patterns depicted in the next illustration because the nearest interseparation distance between every pair is about equal in both cases.

Two distinct patterns not differentiated by the nearest neighbor measure.

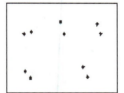
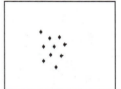

The quadrat analysis of point pattern based on the Poisson process is limited in a different way. Because this method relies solely on frequency counts, the spatial distribution of the quadrats is not considered. As such, this method would fail to distinguish between the two patterns illustrated in the next figure.

Two distinct patterns not differentiated by the method based on the Poisson process.

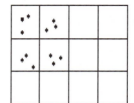 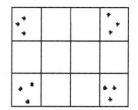

The overall pattern of spatial configuration can be evaluated using spatial autocorrelation statistics. Spatial autocorrelation measures the extent to which the occurrence of one feature is influenced by the distribution of similar features in the adjacent area. As such, spatial autocorrelation statistics provide a useful measure of spatial patterns. If attraction among entities acts as the driving force in the distribution, that is, the existence of one feature attracts similar features to its neighborhood, the spatial autocorrelation is positive and the distribution is characterized by clusters of similar entities.

When competition among features dominates the spatial process—that is, the existence of one entity tends to expel similar entities from its surroundings—, the distribution illustrates a scattered pattern associated with negative spatial autocorrelation. If neither attraction nor repulsion dominates the spatial process, the spatial pattern of distribution is random and no significant spatial autocorrelation exists.

The most commonly used spatial autocorrelation statistic is Moran's I coefficient (Moran, 1948; Cliff and Ord, 1981). This statistic is defined as follows:

$$I = \frac{n\Sigma_i\Sigma_j\delta_{ij}(x_i - \bar{x})(x_j - \bar{x})}{S_o\Sigma_i(x_i - \bar{x})^2}$$

where $S_o = \Sigma_i\Sigma_j\delta_{ij}$. The expected value and variance of this statistic follow:

$$E(I) = -(n - 1)^{-1}$$

$$Var(I) = \frac{n^2 S_1 - nS_2 + 3S_o^2}{S_o^2(n^2 - 1)}$$

where

$$S_1 = (1/2)\Sigma_i\Sigma_j(\delta_{ij} + \delta_{ji})^2$$

and

$$S_2 = \Sigma_i(\Sigma_i\delta_{ij} + \Sigma_j\delta_{ji})^2$$

In the above expression, n is the number of geographic units (points); δ_{ij} denotes the spatial relationship between the i-th and j-th units; x_i denotes the frequency of the spatial phenomenon in question; and S_o is the total number of pairs that hold the spatial relationship.

The value of the I coefficient ranges between -1 and 1. A larger positive value implies a clustered pattern, while a negative value significantly different from 0 is associated with a scattered pattern. When the *I* coefficient is not significantly different from 0, there is no spatial autocorrelation and the spatial pattern is considered random.

The two patterns in the preceding illustration, which cannot be differentiated by the Poisson process, can be clearly differentiated by the *I* coefficient. The distribution at the left shows a relatively clustered pattern with a calculated *I* coefficient of .5, which indicates a positive spatial autocorrelation corresponding to a clustered distribution. The point pattern at the right has a calculated *I* coefficient of -.21, implying a rather scattered distribution.

Point pattern analysis based on spatial autocorrelation can produce inaccuracies, however, because spatial autocorrelation statistics are affected by quadrat resolution. The same pattern of point distribution may generate quite different levels of spatial autocorrelation if the map is overlaid on a different quadrat configuration (Chou, 1995). In some extreme cases, the derived *I* coefficient may not be a valid measure of point distribution. Therefore, when evaluating the spatial arrangement of point patterns using Moran's *I* coefficient, you should consider comparing the coefficient with either the nearest neighbor index (NNI) or the statistics of the Possion process to ensure that the measure is appropriate.

Sampling Point Features

Analysis of point patterns often requires data sampling, especially when the database is large. If spatial properties can be obtained from a sample, processing the entire population is unnecessary and inefficient. On certain occasions, attempting to capture every data point is impossible. For instance, when topography is to be analyzed, surface elevation can be effectively represented by a carefully selected set of data points. In this case, there is no need to try to capture the elevation data for every square inch on the surface.

Another important use of sampling in spatial analysis is the simulation of spatial processes. A hypothetical spatial process may be simulated based on a suitable simulation procedure and an appropriate sampling scheme. In this case, a model must be built to incorporate sampled data into the simulation procedure. Using a GIS, the sampling procedures of point features can be designed in different ways for a comparative analysis. Every sampling procedure is suitable for certain types of applications.

Aspatial Sampling

Sampling can be conducted without reference to spatial components. For instance, assume that you want to carry out a marketing survey of City A. If you wish to obtain a 10% random sample of the population (or at least persons with listed telephone numbers), you could select every tenth record in City A's residential telephone directory and send questionnaires to all selected individuals. Aspatial sampling involves no maps and does not require a GIS. Nevertheless, the sampled data may still be useful in a GIS for spatial analysis. For instance, selected individuals for the marketing survey may be plotted on a map for analysis of spatial distribution, provided that a GIS and adequate street maps are available.

Spatial Sampling

GISs are more useful for spatial sampling in which the sampling procedure is geographically referenced. For instance, to identify households in City A's metropolitan area, you could use census data to identify households in census tracts. Then, you could randomly select 10% of the households in each census tract. Spatial sampling will better represent the geographic distribution of households in a particular census tract than, for example, using telephone listings of households in the tract and selecting every tenth record.

A specified number of point features can be randomly sampled. In the simplest case, all points in the map are consecutively assigned IDs. To sample a point, a random number within the range of IDs is generated and the corresponding point selected. Such a sampling scheme is aspatial because geography is not referenced during the sampling procedure.

Alternatively, a random sample can be taken spatially. For instance, because all points are plotted on a map, a pair of randomly generated x and y coordinates can be used to identify the nearest point feature for sampling. A circle (or another shape) of appropriate size can be randomly plotted on a map, and a point selected if it is the only point falling within the circle. If the circle covers more than one point, all points falling within the circle are selected, or any one of the points is selected.

Both aspatial and spatial sampling can be conducted randomly or systematically. In a systematically stratified sample, certain predefined case selection rules are adopted to meet specific objectives. Consequently, a systematic sample is not a random sample of the entire population in question. For instance, assume that a researcher wants to conduct a survey in a large metropolitan area. A sample of 1,000 households is to be selected from 200 census tracts. A wholly randomized sample generates three or fewer households for some census tracts. The researcher decides that she wants to ensure equal representation of all census tracts in the survey. Her predefined case selection rule becomes "randomly select five cases from every census tract." The resulting stratified sample contains randomly selected cases at the census tract level, but not at the level of the study population (all households in the metropolitan area).

Consider the following instance of aspatial systematic sampling: given that all points are assigned IDs, select every tenth point for a 10% sample. In contrast, a systematic spatial sample could be obtained by placing a grid on a map with a sufficiently large number of cells. All grid cells are systematically assigned IDs. You could randomly pick a number within the range of the IDs to select all points within the chosen cell. A spatially stratified random sample can be taken based on selected polygon units. For instance, if the study area is covered by different types of vegetation polygons, 10 percent of the points within each polygon are sampled.

GISs are especially useful for implementing spatial constraints during sampling. For instance, when point features are spatially autocorrelated, a simple constraint may be specified that no adjacent points should be sampled in order to avoid sampling similar features. In this case, when picking a point, if any adjacent point is already selected, the current point is dropped from the selection set. Another example is that sampled points must satisfy certain spatial criteria, such as within 100 meters of a lake. In this case, buffer zones 100 meters from every lake are generated in the GIS and the sampling applies only to points within the buffer zones.

Spatial Association

Spatial association in point pattern analysis evaluates relationships between point features and other sets of variables. For instance, in the distribution of wildfires, researchers may wish to determine the relationship between ignition site locations and vegetation type.

Spatial association of point features need not be limited to polygon features. For instance, another research question may concern the relationship between ignition site locations and lightning strike point locations. In this case, the spatial association is between two types of points. If the research question pertains to the relationship between points of ignition and roads, then the spatial association would be between points and line features.

Spatial association can be analyzed using regression techniques. The first step is to determine geographic units. Geographic units can be defined in different ways depending on the nature of the problem. In the previous example of the relationship between ignition locations and vegetation cover, geographic units can be defined by ignition points (i.e., every ignition is a record or base unit). In this case, the vegetation type is identified for each point. Alternatively, base units may be defined by vegetation polygons (i.e., each polygon of homogeneous vegetation area is called a unit), and the number of points falling within the boundaries of each polygon is computed.

Summary

This chapter introduced selected methods and respective limitations for the analysis of point features. To ensure reliable results of a point pattern analysis, consider employing more than one method and comparing results. Such an approach requires the use of a GIS for effective manipulation of the data representing the point features.

Point features represent the simplest form of spatial units; distance is the only significant spatial relationship among point features. As such, analysis tends to emphasize spatial patterns in the distribution of point features. If the average distance among a set of point features is small, the distribution would illustrate a concentrated, or clustered, pattern. If the average distance is large, the point features are distributed in a more scattered manner. If a group of point features depicts neither a clustered nor scattered pattern, the underlying process that generates the distribution is considered random.

Point pattern analysis has important implications in a variety of empirical applications. For instance, if the distribution of crime incidents shows a clustered pattern, law enforcement agencies would be able to target areas of concentration for special preventive operations. Crime analysts would be able to build models relating the density of crime incidents to socioeconomic and demographic characteristics. Another example pertinent to ecological systems is that the spatial pattern in the distribution of an animal species usually provides useful clues to a better understanding of the species' natural habitat.

In general, common point pattern analysis procedures include the observation of distributions based on descriptive statistics, examination of spatial arrangement among point features, and further investigation of spatial patterns through evaluating spatial autocorrelation statistics. In addition, GISs are especially useful for researchers in developing spatial sampling schemes.

Exercise

1. According to the attribute tables of point maps A and B below, compare spatial patterns based on nearest neighbor indices (NNIs).

Map A			Map B		
Point-ID	**X-CO**	**Y-CO**	**Point-ID**	**X-CO**	**Y-CO**
1	1	1	1	1	2
2	1	9	2	2	1
3	9	1	3	2	2
4	9	9	4	8	9
5	3	4	5	9	9
6	5	7	6	9	8
7	8	5	7	9	7

A   B

2. Determine whether the three point distributions below are the results of a random process through the use of quadrat analysis of the Poisson process.

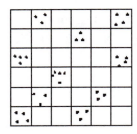
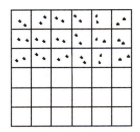

$$P(x) = \frac{e^{-\lambda}\lambda^x}{x!}$$

(x)	O_i	P(x)	E_i	$(O_i - E_i)^2 / E_i$
0		.3679		
1		.3679		
2		.1839		
3		.0613		
4		.0153		
>=5		.0037		

3. Evaluate Moran's I coefficient for the three point distributions shown in question 2. Compare their values.

4. Digitize the point distribution below. Derive the x and y coordinates of all features from the resulting point coverage. The "addxy" function in ARC/INFO is useful for generating the coordinates in the attribute table. In the absence of a GIS, manually estimate the coordinates. Based on the derived coordinates, generate the descriptive statistics of geometric center and standard deviation for both x and y coordinates, and describe the spatial pattern from the statistics. Compute the nearest neighbor index of the distribution and describe the spatial pattern in mathematical terms.

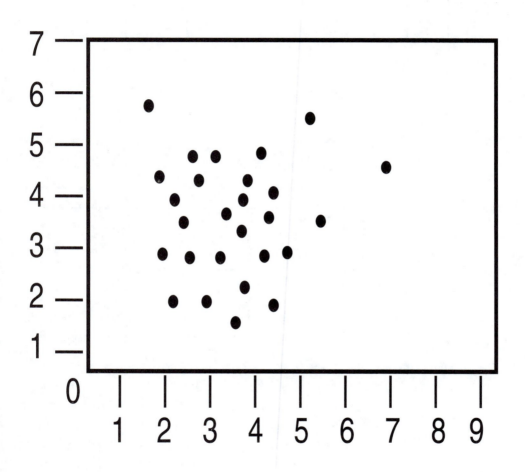

Network Analysis

The central theme of this chapter is the spatial analysis of line features. Line features on maps are usually classified into two major categories, *physical lines* and *virtual lines*. In general, physical line features actually exist in the real world and can be observed from aerial photographs. Rivers, coastlines, and roads are typical examples of physical line features. Virtual line features are intangible such as political or administrative boundaries. Boundaries at various scales between political units are typical examples of virtual line features. Graticules of the geographic grid system (i.e., meridians and parallels) are another type of virtual line feature.

Spatial analysis of line features deals with two types of problems, the structure of connection among line features and the movement in the system through the connected lines. Connected lines define a network, the analysis of which is called network analysis. In most cases, network analysis deals with physical lines such as streets, roads, interstate highways, and the like. Virtual lines do not often affect the mobility or structure of a network.

Traditionally, network analysis is a subdiscipline of transportation research. Topics related to network analysis are covered in transportation geography, urban transportation planning, civil engineering, industrial engineering, and transportation economics. Among the wide variety of transportation problems that can be handled by GISs, this chapter focuses on a few fundamental applications of greatest practical utility to GIS users.

Data Requirements in Network Analysis

A network consists of a number of line segments that are interconnected in some way. Each line segment is defined by start and end nodes, both of known locations. A segment may contain intermediate points of known locations between the start and end nodes. Such intermediate points refine the shape of the segment and are called *vertices*. The main difference between vertices and nodes is that nodes carry information about topological relationships in the network, while vertices exist simply for delineating the segment. The topological relationships defined by the nodes determine the connectivity of a network. The following illustration shows the structure of a typical network resembling a street network structure.

Structure of a typical network.

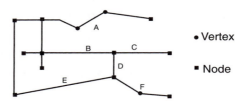

In the preceding illustration, squares represent nodes and circles represent vertices. A line segment may be defined by two nodes. For instance, segments B, C, D, and E have no intermediate vertices. Others may have one or more vertices; segment A has two vertices, while segment F has one vertex. In any case, every segment has two nodes at both ends, a start node and an end node. The diagram illustrates requirements of the network structure in a planar graph, that is, every end of a segment must be a node and every intersection must also be a node. In other words, a planar graph does not allow two line segments to cross each other without creating a node at the intersection.

In addition to the locations of nodes and vertices, each segment is associated with an *impedance* factor, which represents the length or distance from one end to the other. In network analysis, the impedance may be defined according to the nature of the problem. For instance, if travel time is the main concern, then the impedance is defined by travel time, which may be obtained by dividing distance by average speed. Other constraints may be added to the system. For instance, some segments may be directional (e.g., one-way streets). The directional information can be attached to the segment whenever necessary. Other data elements may be useful for specific types of network analysis. For instance, for urban travel problems that involve city streets, special constraints, such as prohibited turns and freeway ramps, may be important in determining the shortest path for delivery routes.

Network topology is a critical element of system specification. Topology is the spatial relationship between objects. In this case, the spatial relationship is embedded in node connectivity. If a segment is directly connected to another, then the two segments must share a common node. If two segments are indirectly connected, then you should be able to traverse through connected segments between these two segments.

Evaluation of Network Structure

γ Index

The structure of a network—relative complexity and connectivity—can be evaluated in several ways. Among available measures of network structure, the γ and α indices measure a network's most fundamental properties. The γ index is defined as the ratio of the actual number of links to the maximum possible number of links in a network. In a planar graph, the maximum number of links is always equal to 3(n - 2), where n denotes the number of nodes. Thus, the γ index is expressed as follows:

$$\gamma = \frac{l}{l_{max}} = \frac{l}{3(n-2)}$$

where *l* is the number of links in a network. The value of the γ index ranges between 0 and 1. A value closer to 0 indicates a simpler network structure with fewer links. A larger value close to 1 indicates a better connected network with more links. For applications based on non-planar graphs, such as those used in airline transportation, the maximum number of links is equal to n(n - 1)/2, instead of 3(n - 2) in the denominator.

The next illustration shows two hypothetical networks where each is defined by eight nodes (n = 8). Network A has a simpler structure with only five links (*l* = 5) while network B has 13 links (*l* = 13). The γ index computed for network A is (5/18 = 0.28), and for network B, (13/18 = 0.72). Thus, network B exhibits the more complex structure.

A

B

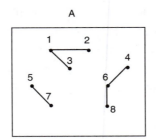

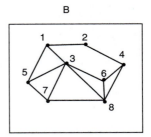

Two hypothetical networks.

α Index

The α index evaluates the structure of a network in a similar way. This index is defined as the ratio of the actual number of circuits to the maximum number of circuits in the network. A circuit is a loop in the network. If three nodes are connected by two links, then there is no circuit among them, and there is only one way to go from any node to any other node. If three links connect three nodes, then one circuit exists because the three nodes are linked by a loop. When a loop exists, each node has two ways to reach any other node in the network.

The α index evaluates network structure in terms of the number of ways you can proceed from one node to another. In order for the α index to be meaningful, all nodes in the network must be connected. If a network is divided into two separate groups of links that are completely separated, the evaluation of circuits is not meaningful. To completely connect a network, the number of links must be at least one less than the number of nodes (i.e., $l = n - 1$). Every additional link beyond that required for the minimally connected network creates a circuit. Thus, the number of circuits in a network is equal to the difference between the actual number of links and the minimum number of links to completely connect a network.

The α index is formally expressed as follows:

$$\alpha = \frac{c}{c_{max}} = \frac{c}{(2n-5)}$$

where c denotes the number of circuits. In the preceding illustration, network A is not completely connected. Consequently, evaluation of the α index is not meaningful for this network. Network B has six circuits; thus its α index is $6/11 = 0.55$.

In general, a better developed transportation network has higher values on both γ and α indices which correspond to higher levels of complexity and connectivity. These indices are useful for evaluating the change in network structure over time or comparing the structures of different networks.

Network Diameter

Diameter is an important measurement of network structure. The diameter of a network represents the maximum number of steps required to move from any node to any other node through the shortest possible routes within a connected network. Network A in the preceding illustration cannot be evaluated for diameter because the network is not connected. In other words, the network is divided into three isolated sets of links and there is no way to go from one set to the other through the network. Network B is a connected network because all nodes are connected, directly or indirectly, through the links in the network. In this case, the diameter of the network is equal to 3 because it takes three steps either to go from node 5 to node 4 or from node 2 to node 7.

Network Connectivity

An important characteristic of a network is how well the nodes in the network are connected. The connectivity of a network can be examined by constructing a matrix set called *C matrices*. In turn, connectivity matrices are useful in evaluating network accessibility.

The first order connectivity matrix, defined as the C^1 matrix, is based on the direct connection between nodes. In brief, if a link exists between a pair of nodes, then the corresponding cell in the C matrix has a value of one; if such link does not exist, the value is zero. Accordingly, the C^1 matrix for network B in the preceding illustration appears below.

C¹ matrix for network B

| | | \multicolumn{8}{c}{**Node number**} | |
		1	**2**	**3**	**4**	**5**	**6**	**7**	**8**	Σ
Node Number	1	0	1	1	0	1	0	0	0	3
	2	1	0	0	1	0	0	0	0	2
	3	1	0	0	0	1	1	1	1	5
	4	0	1	0	0	0	1	0	1	3
	5	1	0	1	0	0	0	1	0	3
	6	0	0	1	1	0	0	0	1	3
	7	0	0	1	0	1	0	0	1	3
	8	0	0	1	1	0	1	1	0	4

Because there are 8 nodes in the network, there are 8 rows and 8 columns in the connectivity matrix. A cell at row P and column Q represents the connection between node P and node Q. Thus, the row 1, column 2 cell has a value of 1 because there is a direct link between node 1 and node 2. For the same reason, the row 1, column 3 cell has a value of 1. The value in the row 1, column 4 cell is 0 because there is no direct link between these nodes.

The last column shows the sum of values for each row. The row sum equals 3 for the first row because there are three ones in the row. The row sum indicates the number of links directly connected to the corresponding node, or node 1 in this case. The largest row sum node is the best direct connectivity node because it is directly connected to the largest number of links. In this example, node 3 is directly connected to five links, and thus, is the maximal direct connectivity node.

Because the C^1 matrix is based on direct connectivity only, it is not sufficient for evaluating network structure. However, transportation systems always consist of both direct and indirect connections. In order to incorporate indirect connections, the C^1 matrix can be extended to a higher order. The product of the C^1 matrix multiplied by itself produces the C^2 matrix, that is, $C^1 \times C^1 = C^2$. Operationally, each cell in the C^2 matrix is the sum of the products of all elements in the corresponding row and column. The row P, column Q cell in the C^2 matrix is the sum of the products of elements in row P and the corresponding elements of column Q in the C^1 matrix.

For instance, the row 2, column 3 cell in the C^2 matrix is the sum of the following eight products: $(1x1)+ (0x0)+ (0x0) + (1x0) + (0x1) + (0x1) +(0x1) + (0x1) = 1$. Every element in the C^2 matrix signifies the number of unique ways that one could move from one node to another through an indirect path of exactly two links. In the row 2, column 3 example, the computed value of 1 indicates that only one way exists to move from node 2 to node 3 in exactly two steps. From the previous illustration, it is clear that the indirect route is from node 2 to node 1, and then from node 1 to node 3. The C^2 matrix of the example network is shown below.

C² matrix for network B

		Node number								
		1	2	3	4	5	6	7	8	Σ
Node number	1	3	0	1	1	1	1	2	1	10
	2	0	2	1	0	1	1	0	1	6
	3	1	1	5	2	2	1	2	2	16
	4	1	0	2	3	0	1	1	1	9
	5	1	1	2	0	3	1	1	2	11
	6	1	1	1	1	1	3	2	2	12
	7	2	0	2	1	1	2	3	1	12
	8	1	1	2	1	2	2	1	4	14

The row sum in the C^2 matrix indicates the number of different ways one could move from the corresponding node to all other nodes in exactly two steps (through an indirect route consisting of two links) in the system. Thus, the sum of the first row is equal to 10, indicating that there are a total of 10 different ways one can move from node 1 to all other nodes in the network in exactly two steps. The row sums of the C^2 matrix indicate that node 3 displays the best indirect connectivity in two steps.

Indirect connectivity can be extended beyond the second order. The product of C^1 and C^2 produces the C^3 matrix, which is the connectivity matrix for indirect connection of exactly three steps (i.e., the indirect routes consisting of three links). The operational procedure for building the C^3 is identical to that for building the C^2 matrix.

In the C^3 matrix, every element represents the number of ways to move from one node to the other in exactly three steps. Thus, the row 2, column 3 cell in this matrix has a value of 3, indicating that there are three different ways one can move from node 2 to node 3 in exactly three steps. The three ways follow: (1) from node 2 through nodes 1 and 5 to node 3; (2) from node 2 through nodes 4 and 6 to node 3; and (3) from node 2 through nodes 4 and 8 to node 3.

C^3 matrix for network B

		Node number								
		1	**2**	**3**	**4**	**5**	**6**	**7**	**8**	**Σ**
Node number	1	2	4	8	2	6	3	3	5	33
	2	4	0	3	4	1	2	3	2	19
	3	8	3	8	4	8	9	9	10	59
	4	2	4	4	2	4	6	3	7	32
	5	6	1	8	4	4	4	7	4	38
	6	3	2	9	6	4	4	4	7	39
	7	3	3	9	3	7	4	4	8	41
	8	5	2	10	7	4	7	8	6	49

Network connectivity should be evaluated in terms of both direct and indirect connections. The first order C matrix represents direct connectivity, while the higher order C matrices represent indirect connectivity of different orders. The evaluation of network connectivity, however, is meaningful only up to the value of the network's diameter. In the previous example, because the diameter of network B is equal to 3, the connectivity matrix should be evaluated up to C^3. Beyond the third order, any additional connectivity matrix contains redundant information.

Network Accessibility

Evaluation of accessibility is an important utility of connectivity matrices. Accessibility in a network can be evaluated in terms of individual nodes or the entire network. In both cases, you must first construct the accessibility matrix, known as the *T matrix*.

The T matrix is the sum of all meaningful C matrices from the first order to the order equal to the network diameter. In the above example, because the diameter is 3, the T matrix is defined as follows:

$$T = C^1 + C^2 + C^3$$

The addition of C matrices takes the form of cell-by-cell addition. In other words, a cell in the T matrix is the sum of values of the corresponding row and column in the C^1, C^2, C^3, ... matrices. The T matrix for network B is shown below.

T Matrix for Network B

		Node number								
		1	**2**	**3**	**4**	**5**	**6**	**7**	**8**	**Σ**
	1	5	5	10	3	8	4	5	6	46
	2	5	2	4	5	2	3	3	3	27
	3	10	4	13	6	11	11	12	13	80
Node number	4	3	5	6	5	4	8	4	9	44
	5	8	2	11	4	7	5	9	6	52
	6	4	3	11	8	5	7	6	10	54
	7	5	3	12	4	9	6	7	10	56
	8	6	3	13	9	6	10	10	10	67
									Σ	426

The row 2, column 3 cell has a value of 4 in the T matrix because the values of the corresponding row and column in the C^1, C^2, and C^3 matrices are 0, 1, and 3, respectively. This value represents the total number of unique ways that one can move directly or indirectly from one node (node 2) to the other (node 3), by making up to two transfers. The row sum indicates the total number of different ways one can move from one node (of the row) to all other nodes through the network. In other words, this value indicates the accessibility of a node. Clearly, a node with a higher row sum has greater accessibility to the rest of the network.

In the current example, the most accessible node in the network is node 3 because its row sum of the T matrix is the highest. A business owner in the real world would be well-advised to locate at Node 3 to maximize accessibility. Node 2, on the contrary, is the least accessible because its row sum is the lowest.

Network accessibility indicates the degree to which one can move in a network from every node to every other node. In this example, the sum of the row sums is equal to 426, which denotes the level of network accessibility. In general, the larger the network accessibility value, the more route choices are available in the system and the better connected the nodes.

Network Structure in a Valued Graph

Network structure evaluation using the γ and α indices and connectivity matrices are based on the simplest definition of line features. Line features are defined most simply by coding the connection between two nodes with a binary variable. If a link exists between two nodes, then the node pair has a connection code of 1; otherwise the code is 0. While useful for evaluating a network's fundamental structure, these methods share a common shortcoming: link representation is oversimplified.

In the real world, any two links are likely to be different lengths; treating them as equal neglects important variations among the links. When the data are available, a network is more accurately represented by a *valued graph*, in which every line segment is coded with a value for a meaningful measure of the link, such as length. In most cases, the value assigned to each link represents an impedance factor based on either the length of the segment, travel time, travel cost, or some combination of such variables.

The next illustration shows a network represented as a valued graph. There are seven nodes (A to G) in the network. Because each link is labeled with a value representing the link, the structure of the network can be evaluated based on the values.

A network represented as a minimal tree valued graph.

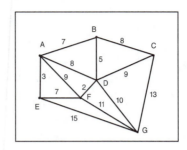

Minimal Spanning Tree

A minimal spanning tree, one type of valued graph, is a specific network that satisfies three criteria. First, the tree connects all nodes in the network with a minimal number of links. This criterion requires that the number of links is one less than the number of nodes, that is, $l = n - 1$, where l denotes the number of links and n denotes the number of nodes. Second, the root of every tree is located at one of the nodes in the network. Accordingly, the number of trees that can be constructed is the same as the number of nodes in the network. Third, the distance between each node and the root of the tree is minimized.

A minimal tree rooted at any node can be constructed for the network, as shown in the previous illustration. Construction of the minimal tree requires six iterations because there are six nodes to reach from every node. The following procedure explains how the minimal tree rooted at node A is built.

1. Find the link of minimum cost (for distance) extending from node A. In this example, A to E is the minimal cost link (3) among all links directly connected to node A. Label the minimum cost link (A-E), and record the node at the other end of the link (E). Next, record the cost at node E, which is 3.

2. Repeat the above procedure, this time extending from both nodes A and E, because E has been reached. It is important to note that an extension from node E must start with a nodal cost of 3, while an extension from node A has a nodal cost of 0. Thus, the cost for reaching node F directly from node A is 9, while the cost for reaching the same node from node E is 3 + 7 = 10.

3. Identify the link of minimum cost to reach. Label the link, the reached node, and the cost of reaching that node. In this case, A-B is the link, and B is the reached node with a cost of 7.

4. Repeat the procedure for six iterations. A node is reached and a link is labeled every time. Thus, in every iteration the tree grows from the root by one additional branch (link). This procedure ensures that the cost to reach any node from node A is of minimum cost.

The next illustration shows the results of the six iterations; the nodal cost for the reached node at each iteration is indicated in parentheses. The nodal costs of all nodes in the network, implying the minimum costs from node A, can be organized into an array in alphabetical order of the nodes, that is, [0, 7,

15, 8, 3, 9, 18] for nodes A to G. The sum of all cells in this nodal cost array is equal to 60, which is the total network cost for the tree rooted at node A. The total network cost is the minimum total cost for moving from A to every other node in the system. A smaller total cost number indicates a more centralized location of the root node.

Procedure for constructing a minimal spanning tree rooted at node A.

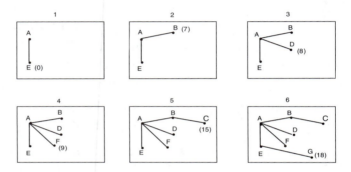

The next illustration shows two minimal trees rooted at nodes D and G, respectively. The cost array for the tree rooted at D is [8, 5, 9, 0, 9, 2, 10] with a total network cost of 43. The cost array for the tree rooted at node G is [18, 15, 13, 10, 15, 11, 0] with a total network cost of 82. Clearly, the tree rooted at D is much more efficient than the one rooted at G, indicating that D has a more centralized location than G.

Two minimal spanning trees with different roots.

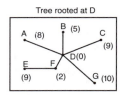

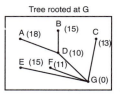

A typical application of minimal spanning trees is the identification of street segments within a specified distance from an existing facility. For instance, in school bus routing, it is necessary to identify students who live too close to respective schools to receive free bussing services. The next diagram shows a GIS routing application for identifying street segments within a specified distance from a school (e.g., 0.5 miles). The square symbol in the middle of the diagram represents the school's location.

A GIS is used to select all street segments (highlighted) within the specified "walking distance" from the school. Street segments are selected within the specified distance by extending from the school minimal spanning trees in all directions until the distance criterion is satisfied. A simple database manipulation matching student addresses to street addresses will then identify all students who are ineligible for bussing services because they live on the selected street segments. This procedure enables school administrators to effectively separate students eligible for bussing services from those who are not eligible.

Street segments within a specified distance from a school can be identified through a minimal spanning tree.

Shortest Path Algorithm

A typical problem in network analysis is finding the shortest path from one node to another through a network. Because all data in GISs are organized in digital format, such questions can be effectively answered using developed algorithms. An algorithm is a set of mathematical expressions or logical rules that can be followed repeatedly to find the solution to a question. For the shortest path problem, a simple algorithm utilizing a procedure similar to the construction of minimal spanning trees can be adopted to find the optimal solution.

To find the shortest path from the origin node to the destination node, the analyst follows the procedure of tree building rooted at the origin node. The procedure starts with finding the first, nearest node from the origin node, and then continues with finding the next nearest node. In each iteration, the tree grows another branch. As soon as a node is reached as the result of this additional branch, a logical condition is evaluated to determine if the reached node is the destination node. If the condition fails, the procedure goes on to the next iteration. When the destination is reached, the algorithm can be terminated because the shortest path has been found.

To find the shortest path from node A to node G, for instance, the tree is built rooted at A; the solution is either from A through E to G or from A through D to G at a cost of 18. To find the shortest path between D and E, the tree must be rooted at D and the solution, according to the preceding illustration, is from D through F to E at a cost of 9.

Minimal spanning tree rooted at node A.

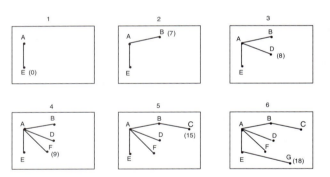

The shortest path algorithm is useful for a wide variety of transportation problems. Although this algorithm is used to find the shortest path from any given origin to any destination, the routing solution can be expanded to solve complex problems with a combination of origin, multiple intermediate stops, and destination. Applications include routing school buses with known bus stops and routing emergency vehicles such as ambulances or fire engines, among many others.

Traveling Salesman Problem

An important and practical application of network analysis is to find the optimal route for trips that involve multiple stops. The optimal solution for the "traveling salesman problem" requires a salesman to travel from an origin node to several sites of known location. The required data include the locations of the origin, multiple destinations, and the impedance factor (e.g., travel cost) between every pair of nodes. The objective is to find the route for which the total impedance (e.g., total travel cost) is minimized.

The algorithm for solving the traveling salesman problem is explained using the following hypothetical example. A publishing company delivers magazines to six book stores located at nodes A through F in the next figure. These locations are connected by a freeway network with the distances labeled. The task is to find the optimal delivery route with the least possible total travel distance.

Finding the optimal delivery route on a freeway network.

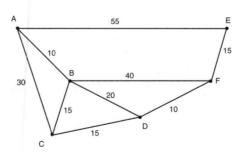

The first step is the construction of a cost matrix among the locations. The cost matrix below contains as elements the cost values extracted from the network above. Each cell represents the distance between the corresponding location pair. When there is no direct link between a pair of locations, the distance is indicated by a minus sign (-).

Iteration 1

Cost matrix 1

		A	B	C	D	E	F	Min
	A	-	10	30	-	55	-	10
	B	10	-	15	20	-	40	10
Node	C	30	15	-	15	-	-	15
	D	-	20	15	-	-	10	10
	E	55	-	-	-	-	15	15
	F	-	40	-	10	15	-	10

Each iteration requires a procedure comprised of the following four steps: (1) a row minimum operation, (2) a column minimum operation, (3) penalty assignment, and (4) path identification. The purpose of the row minimum operation is to identify the minimum cost of each row and then subtract the minimum from every cell of the row. This step ensures that every row has at least one cell with a value of 0. In cost matrix 1, the row minimum is identified and then subtracted from every cell of the row to generate cost matrix 2.

Cost Matrix 2

		A	B	C	D	E	F
				Node			
	A	-	0	20	-	45	-
	B	0	-	5	10	-	30
Node	C	15	0	-	0	-	-
	D	-	10	5	-	-	0
	E	40	-	-	-	-	0
	F	-	30	-	0	5	-
	min	0	0	5	0	5	0

The second step is to identify the column minimum and then subtract the minimum from every cell in the corresponding column. This step ensures that every column has at least one cell with a value of 0. The result of the column operation is cost matrix 3.

Cost matrix 3

		A	B	C	D	E	F
				Node			
	A	-	0	15	-	40	-
	B	0	-	0	10	-	30
Node	C	15	0	-	0	-	-
	D	-	10	0	-	-	0
	E	40	-	-	-	-	0
	F	-	30	-	0	0	-

In cost matrix 3, there is at least one 0 cell in every row and column. This matrix is now ready for penalty assignment. The penalty is to be assigned to only those cells with a value 0. Other cells need not be considered in this step. For every candidate cell with a value of 0, the penalty consists of a row component and a column component. The row component is the minimum value in the same row, excluding the cell under evaluation. Likewise, the column component is the minimum value in the column excluding the cell under evaluation. The penalty is the sum of the two components. The assignment of penalty based on cost matrix 3 is shown in the penalty matrix below. In this matrix, the value in parentheses in each candidate cell indicates the penalty assigned to that cell. For instance, the 0-value cell from location A to location B is the sum of the minimum row value (15) and the minimum column value (0), or (15 + 0 = 15).

Penalty matrix of iteration 1

		Node					
		A	**B**	**C**	**D**	**E**	**F**
Node	A	-	$0^{(15)}$	15	-	40	-
	B	$0^{(15)}$	-	$0^{(0)}$	10	-	30
	C	15	$0^{(0)}$	-	$0^{(0)}$	-	-
	D	-	10	$0^{(0)}$	-	-	$0^{(0)}$
	E	40	-	-	-	-	$0^{(40)}$
	F	-	30	-	$0^{(0)}$	$0^{(40)}$	-

According to the penalty assignment, the maximum penalty candidate is the most critical link for the minimum cost route. When a tie exists, any maximum penalty candidate link can be selected for inclusion in the route. In this example, penalty matrix 1 indicates that either the link E-F or the link F-E is the critical link and must be included in the route. Thus, the link E-F is selected in this iteration. In other words, each iteration selects a link to be included in the optimal route; consequently, the number of iterations required to build the optimal route is one less than the number of locations in question.

At the end of each iteration, the cost matrix must be revised by (1) eliminating the row and column of the location pair corresponding to the selected link, and (2) revising to infinity the cost of the link opposite the currently selected link in order to rule out the possibility of a return trip on the same link. Cost matrix revision in each iteration ensures that the trip will not be duplicated in an infinite loop. The revised matrix is presented in cost matrix 4.

Cost matrix 4

		A	**B**	**C**	**D**	**E**
				Node		
Node	A	-	0	15	-	40
	B	0	-	0	10	-
	C	15	0	-	0	-
	D	-	10	0	-	-
	F	-	30	-	0	-

Iteration 2

In cost matrix 4, every row and every column has a 0-value cell except for column E. Consequently, the row operation is not necessary and only the column operation is required. The resulting cost matrix and the assigned penalty for each 0-value candidate are shown in the penalty matrix.

Penalty matrix of iteration 2

		Node				
		A	**B**	**C**	**D**	**E**
Node	A	-	$0^{(0)}$	15	-	$0^{(-)}$
	B	$0^{(15)}$	-	$0^{(0)}$	10	-
	C	15	$0^{(0)}$	-	$0^{(0)}$	-
	D	-	10	$0^{(10)}$	-	-
	F	-	30	-	$0^{(30)}$	-

This iteration selects the A-E link according to the penalty matrix. Adding this link results in a route sequence of A-E-F, which means from location A to F via E. The revised cost matrix is presented as cost matrix 5.

Cost matrix 5

		Node			
		A	**B**	**C**	**D**
Node	B	0	-	0	10
	C	15	0	-	0
	D	-	10	0	-
	F	-	30	-	0

Iteration 3

In cost matrix 5, because every row and column has at least one cell with a 0 value, there is no need to conduct either row or column operations. The corresponding penalty assignment is exhibited.

Penalty assignment

		Node			
		A	**B**	**C**	**D**
Node	B	$0^{(15)}$	-	$0^{(0)}$	10
	C	15	$0^{(10)}$	-	$0^{(0)}$
	D	-	10	$0^{(10)}$	-
	F	-	30	-	$0^{(30)}$

In the preceding iteration, the candidate of maximum penalty is the F-D link. This link is therefore included in the route. The route sequence now contains three links, A-E, E-F, and F-D. Because the F-D link is included, revision of the cost matrix eliminates row F and column D.

Iteration 4

The cost matrix is revised at the end of the preceding iteration, and the resulting penalty assignments are shown below.

Penalty assignment

		A	B	C
Node	B	$0^{(15)}$	-	$0^{(0)}$
	C	15	$0^{(25)}$	-
	D	-	10	$0^{(10)}$

In the above iteration, the maximum penalty link is C-B. This link is not connected to the current route sequence, A-E-F-D. Thus, after selecting this link, the current route consists of two separate route sequences, A-E-F-D and C-B. The revision of the cost matrix involves the elimination of row C and column B, and the alteration of the cost for the link B-C to non-existence.

Iteration 5

Penalty Assignment

		A	C
Node	B	$0^{(-)}$	$0^{(0)}$
	D	-	$0^{(-)}$

The revised cost matrix and the resulting penalty assignments are shown in the table above. In this case, two of the candidates, B-A and D-C, have a value considered to be infinity. If the former is selected for inclusion in the route, then a route connecting the two separate sequences is identified, that is, C-B-A-E-F-D. At this point, all six locations are included in the route sequence, and a return link from D to C completes the route to the site of origin. Because there are six locations in the problem, the complete route sequence is obtained in five iterations. This concludes the traveling salesman algorithm.

Construction of delivery networks is a typical application of the traveling salesman algorithm. In this case, the algorithm is modified with additional local checks to deal with multiple origins and receiving locations. Receiving locations are clustered and then assigned to respective closest sources. An optimal route is determined for each cluster. The objective is to find the optimal routing solution with a minimum total travel cost. Although every route is a traveling salesman problem, the problem of delivery networks involves finding the best combination of receiving stops for each source.

The next diagram shows the street network for a part of Orange county in southern California. The street data are derived from the U.S. Census Bureau's TIGER/Line file. The upper left triangle denotes the start location of a route. This start location could be the transportation yard where vehicles are housed. The circles represent three stops which the route must include. In many cases, stops must be visited in a certain sequence.

Street network, Orange county (southern California) showing route and stops.

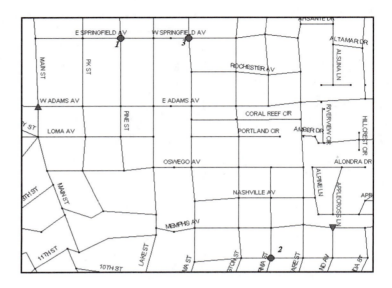

For instance, a vehicle must first go to stop 1 to pick up advertising materials from a client. At stop 2, a printing shop, multiple copies of the ad are printed. Next, the driver must deliver prints to the post office, stop 3, to mail out the materials. Finally, the vehicle is to return to the yard, the end location indicated by an upside down triangle. When the stops must be visited in a specific order, a shortest path is generated between each origin-destination pair. The most effective routing solution using a GIS is shown in the next diagram. The solution is derived by directly applying the shortest path algorithm to the street map. Note that arrows on street segments indicate route directions. Double arrows (in both directions) imply segments that would be traveled twice in the opposite directions.

*Most effective routing
solution derived through
use of a GIS.*

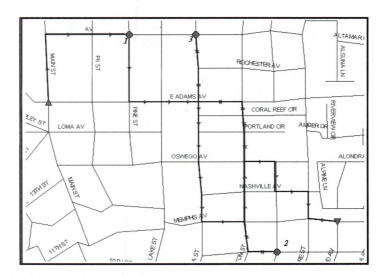

In addition to the map display of the most effective route according to the routing requirements, a table listing the travel direction, driving time, and distance on each street can be generated for drivers who are not familiar with the area. The following diagram shows part of a routing directions table.

Routing directions table.

Street	Turns	Miles	Minutes	Loadi
#Start:			0	
MAIN ST		0.1383	0.1844	
E SPRINGFIELD AV	Right	0.1667	0.7858	
#2:			5	
PINE ST		0.1382	0.2369	
W ADAMS AV	Left	0.0633	1.0843	
E ADAMS AV	Straight	0.0748	0.0537	
ALABAMA ST	Right	0.249	0.9269	
MEMPHIS AV	Left	0.0863	1.1479	
HUNTINGTON ST	Right	0.0624	0.6071	
LINCOLN AV	Left	0.0575	1.0986	
#3:			5	
LINCOLN AV		0.0575	0.0986	
HUNTINGTON ST	Right	0.3115	1.0339	
E ADAMS AV	Left	0.0863	1.115	
ALABAMA ST	Right	0.1384	0.7373	
#4:			5	
ALABAMA ST		0.1384	0.2373	
E ADAMS AV	Left	0.0863	1.115	
HUNTINGTON ST	Right	0.1247	0.7136	
OSWEGO AV	Left	0.0623	1.1068	
CALIFORNIA ST	Right	0.0622	0.6066	

On other occasions, routing among several stops may not need to follow a specific order. For instance, the delivery of pizza or furniture may require route efficiency only, rather than a particular sequence of stops. In this case, routing efficiency can be defined by the total travel distance.

In the next illustration, assume that a messenger carrying legal notifications leaves the site (court house) indicated by the triangle, delivers the notifications to three individuals (circles) involved in a case, and then returns the vehicle to a maintenance yard (upside down triangle). This scenario illustrates a typical traveling salesman problem. The GIS-generated solution is also shown in the diagram below. Note that the route is different from that of the previous example, in which a particular sequence of stops was required.

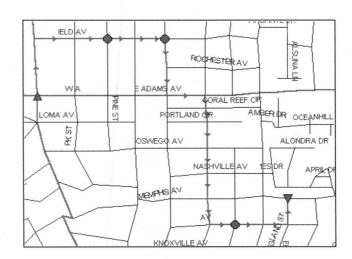

GIS-generated solution to typical traveling salesman problem.

Shipment Problem

Shipment problems deal with optimizing the transportation of goods or people from multiple origins to multiple destinations. Once again, the objective in terms of optimization is to minimize the total transportation cost. The optimal shipment schedule is to be determined based on the following conditions.

❒ The supply quantity at each of several source locations and demand quantity at each of multiple receiving locations are known.

❏ The transportation cost between each source location and each receiving location is known.

❏ In simpler cases, it is usually assumed that total supply equals total demand, although this assumption may be relaxed for more realistic situations.

❏ Routing is not a concern in the shipment problem.

The following example is a typical shipment problem. Assume that an automobile corporation has three component manufacturing plants, A, B, and C, and three assembly plants, I, J, and K. The next table shows the transportation cost between each component plant origin and assembly plant destination. For instance, the transportation cost for each unit shipped from component plant A to assembly plant I is $8.

Transport costs ($/unit)

		Assembly plant		
		I	**J**	**K**
Component plant	A	8	3	6
	B	7	11	5
	C	6	2	10

The table below lists the current shipment schedule, which is not optimized. The 0 value from plant A to plant I indicates that no shipments between these plants are scheduled. Next, the value of 10 from A to J indicates that 10 units are shipped from plant A to plant J every shipment period. The sum of each row shows the supply quantity from the corresponding component plant. For instance, the supply quantity is equal to 10 for plant A. The sum of each column denotes the demand quantity at the corresponding assembly plant. For instance, the demand quantity is 7 for assembly plant I. The task at hand is to alter (optimize) the shipment schedule by minimizing total transport cost.

Current Shipment Schedule

		Assembly plant			
		I	**J**	**K**	**Σ**
Component plant	A	0	10	0	10
	B	0	0	6	6
	C	7	1	6	14
	Σ	**7**	**11**	**12**	**30**

The first step toward optimization of a shipment schedule is to determine whether the existing schedule is optimized. This step is accomplished by evaluating the current shipment schedule with the assumption that the system is optimized. Shipment schedule evaluation is based on *shadow prices* at all origin and destination sites. A shadow price represents the relative price of a commodity at a certain locality; it is not the real price. Local prices may vary from place to place due to numerous factors, while shadow prices for different locations are different because of transportation costs.

When a system is optimized, and assuming the local price for purchasing one unit of a given commodity at the site of origin is known, then the local price at the destination site must be equal to the price at origin plus transportation costs. If the difference in the origin and destination site prices is less than transport cost, then the shipment system is not optimized in that every unit shipped from the origin to the destination site results in a loss. In contrast, if the difference in the origin and site prices is greater than transport cost, the shipment schedule is still not optimized because additional units should be shipped from the origin to destination to take full advantage of the price differences. Shadow prices can be calculated for every origin-destination site shipment based on the above concept.

Because shadow prices represent local prices expressed in relative terms, an arbitrary value can be assigned to any site to begin evaluating shadow prices. In this example, $0 is assigned to the first site of origin (component plant A). According to the current shipment schedule, there is only one shipment from plant A, which consists of the 10 units shipped to plant J. Thus, if the assigned price is $0 at plant A, then the unit price at the assembly plant J must be $3, because the unit transportation cost from A to J is $3.

The shadow price for plants I and K cannot be determined because there is no shipment between this origin-destination pair. However, because the shadow price is $3 at plant J, and there is a shipment of one unit from plant C to plant J, the price must be $1 at C because the transport cost is $2 between C and J. Shadow prices at other sites can be determined in the same way. The estimated shadow prices at the component plants are $0, $6, and $1 for A, B, and C, respectively. The shadow prices for the assembly plants are $7, $3, and $11 for I, J, and K, respectively.

Shadow prices

Origin	A	B	C
	0	6	1
Destination	**I**	**J**	**K**
	7	3	11

When all shadow prices are estimated, you can determine whether the current shipment schedule is optimized by evaluating the origin-destination pairs for which there are no shipments. The shadow prices are evaluated by assuming that the system is optimal according to the price differences between origins and destinations. The evaluation of shadow prices is based solely on origin-destination pairs with existing shipments. If the price difference is larger than transportation cost for pairs without shipments, there is a potential profit which is not currently realized. Whenever such a potential profit exists, the system is apparently not optimal.

In the current shipment schedule, there are four origin-destination pairs without shipments, A-I, A-K, B-I, and B-J. The price difference between A and I is $7 because the shadow price at A is $0 and at I, $7. The transportation cost is $8. Thus, the potential profit for any shipment between A and I is -$1. A negative profit means that a loss will occur if shipments take place between the corresponding origin-destination pair.

Potential profit

	I	J	K
A	1	0	5
B	-6	-14	0
C	0	0	0

The price difference between B and I is $1, while the transportation cost is $7; consequently, the potential profit for any shipment from B to I is -$6. The price difference between B and J is -$3, and the transportation cost is $11, for a potential profit of -$14 for the B-J pair.

Shipments between the A-I, B-I, and B-J pairs will not produce potential profit. However, the price difference between plant A and plant K is $11, while the transportation cost is only $6, implying a potential profit of $5. The existence of a positive potential profit indicates that the current system is not optimal and additional unit shipments should take place between plants A and K.

There are many ways in which shipments between A and K could be increased. However, to do so without affecting the balance of supply and demand requires identifying other origin-destination pairs for which a shipment reduction is possible. Ideally, shipment rescheduling should reduce the shipment between a origin-destination pair with a relatively high transportation cost. In this example, the transport cost from plant C to K is the highest ($10). Consequently, shipment rescheduling can be executed by reducing the current 6 units shipped from C to K to 0 units.

The same unit amount (6) can be added to the A-K pair without affecting total demand at plant K. The additional 6 units from A to K can only derive from the current 10-unit shipment between A and J. Thus, shipment for the A-J pair must be reduced by 6 units to 4 units. Likewise, the reduction of 6 units shipped from C to K must instead be shipped from C to J; the C-J pair shipment becomes 7 units. The adjusted shipment schedule is shown in the following table. Note that the shipment rescheduling does not affect the row sums (supply quantities) or column sums (demand quantities).

Revised shipment schedule

	I	J	K	Σ
A	0	4	6	10
B	0	0	6	6
C	7	7	0	14
Σ	**7**	**11**	**12**	**30**

Shadow prices for the component plants, computed for the revised shipment schedule, are $0, $1, and $1 for A, B, and C, respectively. Prices for assembly plants are $7, $3, and $6 for I, J, and K, respectively. Potential profit must be evaluated for the A-I, B-I, B-J, and C-K pairs. When following the procedure described above, potential profits are obtained for the four pairs of -$1, -$1, -$9, and -$5, respectively. Because there is no more positive potential profit, the revised shipment schedule has been optimized. If, at this stage, positive potential profit existed, then additional iterations of the algorithm must be executed until no more positive potential profit exists.

The method discussed in this section is useful for optimizing shipment schedules in a variety of transportation applications that involve multiple sources and multiple delivery locations. The algorithm can be modified for cases with special constraints. For instance, if the total supply is more than the total demand, the supply surplus can simply be allocated to a storage site.

Network analysis—the spatial analysis of line features—is useful for a wide variety of transportation and location problems. This chapter introduced selected fundamental methods that are used to evaluate the structure of a network, find the optimal route in a network, and optimize shipment schedules for typical multiple origin, multiple destination problems.

A variety of methods are available for more complicated problems with additional constraints. For instance, a special, yet widely employed type of transportation network is the hub-and-spoke system. A hub-and-spoke network consists of one or more centrally located nodes that serve as hub facilities for the network and numerous spoke nodes. Hub-and-spoke networks are found in airline transportation and express package delivery services, where a node-to-node network is inefficient. In the case of airline transportation, major carriers operate on hub-and-spoke networks in order to minimize operating costs. Many passengers traveling between spoke locations must make transfers at one or more hub locations.

The following illustration shows the optimal structure of a double-hub air transport network for the 30 top airline traffic cities in the U.S. for 1985. In this system, the hubs are located in Cleveland and Denver. The structure is optimal in the sense that the total travel cost for all passenger traffic among the 30 cities is at a minimum.

Optimal structure of a hub-and-spoke network with hubs located in Cleveland and Denver, 1985.

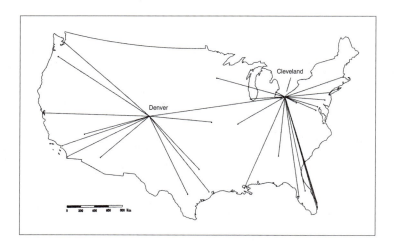

In the next illustration, St. Louis and Las Vegas are the optimal locations for a double-hub network serving the same 30 cities based on 1989 airline traffic. The move westward between 1985 and 1989 accommodates the higher demand growth for air transportation on the west coast.

St. Louis and Las Vegas, optimal locations of hub facilities in 1989.

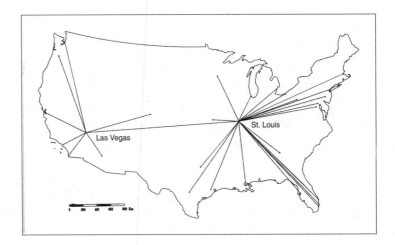

The following illustration shows the proportional change in air transportation accessibility for flights with up to two transfers in the United States between 1970 and 1980. The spatial pattern indicates a rapid increase in air traffic in the west and southwest, while New England and the Great Lakes area experienced a significant decline in air transportation accessibility.

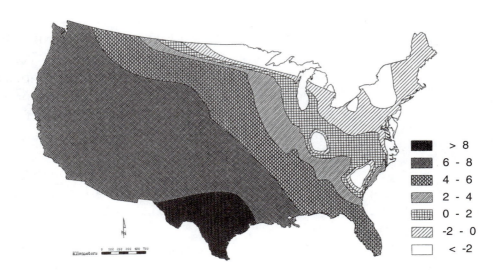

Proportional change in air transportation accessibility for flights with up to two transfers in the United States between 1970 and 1980.

Summary

Many GIS-based network analysis applications are available. Certain forms of transportation, commuting, shipment, delivery, communication, and routing are part of almost everyone's daily life. Digital data of line features, such as roads, streets, freeways, and communication networks, are now among the most widely used sources of spatial information. Fundamental issues pertaining to the spatial analysis of line features comprised the central theme of this chapter.

Analysis of a transportation system or a communication network starts with a careful evaluation of network structure. Evaluation of network structure allows the analyst to understand the potentials and limitations of the system under consideration. The most critical element of information about spatial relationships among a set of line features is connectivity, that is, how well the network is connected. Indeed, analysis of network connectivity is a prerequisite to effective design and management of transportation systems. For instance, calculating the cost-effectiveness of new construction versus expansion of the existing street network can only be evaluated by means of a thorough investigation of the impact of each alternative proposal.

For routing applications, transit agencies need to design regular routes through a series of network analyses. This chapter introduced two of the most commonly employed indices for evaluating network structure. The C matrix of network connectivity and the T matrix of network accessibility can be evaluated using different methods. C matrices of network connectivity indicate the degree to which nodes and links are connected in a transportation system. T matrices indicate system users' level of access to transportation services.

Important transportation applications of network analysis include (1) building the shortest path between any specified sites of origin and destination, (2) the traveling salesman scenario which allows for a set of stops (locations) to be visited in the most efficient way, and (3) the shipment problem which allocates shipments within a transportation system while minimizing total transport costs. With the inherent capability of network analysis, along with relatively low-cost street data such as the TIGER/Line file by the U.S. Census Bureau, modern GIS technology provides a powerful tool for a wide variety of organizations, including trucking companies, grocery chains, delivery services, courier services, airport shuttle services, school districts, fire departments, and emergency medical services, among many others.

Exercise

1. Six commercial districts (A to F) are connected by a highway network depicted below. Two transportation improvement projects have been proposed. The first proposal will build a highway connecting A and D while the second proposal will link B and E. The construction costs are identical for both proposals.

(a) Build the connectivity and accessibility matrices of both proposals.

(b) If the first proposal is implemented, which district will benefit the most in terms of direct connectivity? Which district will gain the greatest advantage in network accessibility? Answer the same questions for the second proposal.

(c) Which proposal would you recommend if the objective is to improve overall transportation efficiency (benefits) for the entire system?

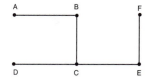

2. A retail chain has three production centers (A, B, and C) and three distribution centers (I, J, and K). The supply-demand schedule, unit transport costs, and the existing shipment schedule are given in the following tables. Find the optimal shipment schedule with a minimum total transport cost, and identify the percentage unit saving in transport cost.

Supply-demand (000 units)

Production centers	Supply	Distribution centers	Demand
A	10	I	12
B	20	J	16
C	15	K	17

Transport costs ($/unit)

	I	J	K
A	6	4	3
B	5	3	8
C	2	6	5

Current shipments (000)

	I	J	K
A	10	0	0
B	2	16	2
C	0	0	15

3. Six neighborhoods are connected by a highway network depicted below. The number attached to each link denotes travel distance. A medical center is to be located in one of the neighborhoods. The objective is to locate the facility in such a way that the total travel distance from the facility to all neighborhoods is minimized. In order to identify the optimal location, you must construct a minimal spanning tree rooted in each neighborhood, compute the total travel distance, and find the system of minimum total travel distance.

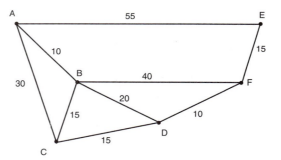

4. An automobile corporation is seeking the optimal location to build a new plant. The corporation has six component factories, A to F. The diagram below shows locations and the highways connecting them. To take advantage of existing facility, the proposed plant will be located at one of the six existing factories. Production at the proposed plant will require the use of components provided by every factory. Component weights from the six factories are estimated to be 1.0, 0.8, 1.2, 1.5, 2.0, and 0.5 tons for factories A through F, respectively. Your task is to locate the proposed plant while minimizing the total transport cost.

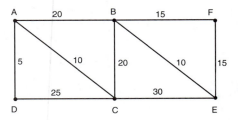

Spatial Modeling

This chapter focuses on constructing models for analyzing spatial phenomena. Spatial modeling serves the following two related purposes.

1. A spatial model analyzes phenomena by identifying explanatory variables that are significant to the distribution of the phenomenon and providing information about the relative weight of each variable.

2. Spatial models are most useful for predicting the probable impact of a potential change in control factors. In other words, spatial models can be used to estimate the results if the values of selected variables are altered.

For instance, a researcher might analyze soil erosion variance by studying a large number of environmental variables, including slope gradient, slope type, vegetation coverage and the like. A comprehensive database may include numerous environmental variables. Some variables may play a critical role in soil erosion, some may be much less significant, while others may not be significant at all. Next, correlation among variables is not uncommon. Incorporating correlated variables into a spatial model induces inflated parameter variances and should be avoided. In spatial modeling, the statistical significance of each variable can be tested in order to determine the most effective model with a minimum set of explanatory variables. Moreover, the estimated coefficient can be used as a measure of the relative weight of each variable.

The second purpose of spatial modeling answers "what-if" questions by evaluating alternative hypothetical situations. Possible situations that do not exist at the time of analysis can be generated by assigning values to meaningful variables and calibrating the model. Distributions of the study phenomenon based on hypothetical scenarios can then be mapped for examination of potential effects.

Spatial modeling is applicable to all feature types (points, lines, and polygons). Because polygon features are the most complicated form of the three feature types, the discussion in this chapter focuses on polygon features. Methods for building spatial models of polygon features can be applied to point features or line features as well.

Model Building

When building a spatial model to analyze the distribution of a phenomenon, the analyst must consider several questions, including the nature of the model, the definition of geographic units, the specification of explanatory variables, and the quantitative method to employ. Spatial models are generally divided into four types according to the primary purpose of the model: descriptive, explanatory, predictive, and normative. The process of model building can also be classified into four stages, corresponding to model types. However, it is difficult to separate modeling stages because they are often interdependent rather than consecutive. After all, it is impossible to explain the distribution of any phenomenon unless one can describe the distribution, and it is impossible to predict a spatial pattern unless one can explain the effects of significant factors.

Description

In the first stage of modeling, the principal objective is to characterize the distribution of the spatial phenomenon. In general, description deals with spatial order and is represented by a set of descriptive statistics and indices. For instance, questions about wildland fires, such as the mean size of the fires and their distribution throughout a landscape or concentrated in certain areas, are in the domain of descriptive models.

In general, descriptive models provide the basis and determine the need for other stages of model building. If the distribution shows a systematic pattern, it is then necessary to explain it by identifying factors related to the systematic pattern in the distribution. If, on the contrary, the distribution exhibits a random pattern, which suggests that the realization is the outcome of a random process, building an explanatory model may not be necessary.

Explanation

Explanation, the second stage of model building, deals with spatial association, or relationships between the phenomenon and the factors affecting the distribution. For instance, in order to explain why wildfires tend to occur in certain areas, roles of vegetation, topography, climatic factors, human activities, and so on, must be examined. The key issue is how each factor influences fire occurrence.

Prediction

Once the distribution of a spatial phenomenon is explained, predictive models can be constructed. Predictive models are most useful for two purposes: prediction, and simulation of alternative management strategies. The first type, given current conditions and growth factors of significant variables, predictive models provide description of the phenomenon in question in a subsequent time period. Alternative management strategy simulation asks what will happen if certain conditions are altered. These problems can be handled by calibrating the models, that is, plugging in known parameters of the explanatory models and deriving results of the distribution.

Normative Modeling

Constructing normative models that provide optimal solutions is the most complex type of model building. This activity typically requires different types of modeling techniques such as linear or nonlinear programming. Normative models are especially useful for planning and land management, where an existing system must be improved for increased efficiency.

The general procedure for building a normative model includes the specification of an objective function and the identification of constraints. In a typical transportation planning problem, the objective function is to minimize total transport cost, which is subject to constraints. For instance, a company might need to equalize the total volume shipping out from production sites and the total received volume at all destinations.

While it is possible to incorporate optimization methods for spatial modeling, in spatial analysis most applications of normative models are limited to line features in transportation networks, including transportation applications and location analysis. Some of the methods introduced in Chapter 7 are employed in normative modeling for transportation applications. The discussion of spatial modeling in this chapter is limited to the first three types of models: descriptive, explanatory, and predictive.

Delineation of Geographic Units

The first step toward spatial model construction is the delineation of geographic units. Appearing below is a summary of how geographic units may be defined.

Arbitrary Boundary Delineation

Geographic units defined by arbitrarily delineated boundaries usually satisfy certain modeling criteria. The most common criteria include simplicity of shape, homogeneity of area, and ease of geocoding and referencing. A grid is a typical example in which every grid cell represents a geographic unit for analysis.

An important advantage of a grid-defined system is that all geographic units are regularly spaced, and thus, spatial relationships are fixed and easily traceable. Furthermore, as variation in geometric properties is minimized, spatial models based on a grid system are not subject to area adjustment.

There are also several disadvantages to the arbitrarily defined system. Existing data may not be available for grid cells and thus, data conversion is required. Next, because the boundaries are arbitrarily defined, they do not reflect any useful boundary delineation in practical use. Consequently, direct interpretation of results may be difficult. In addition, modeling results may be unstable over the resolution and placement of the grid.

Units Delineated by Existing Boundaries Created for Other Purposes

Existing boundaries are defined either for administrative purposes, or according to political units or historical divisions. In many cases, administrative or political units define the fundamental units of available spatial data. Spatial models for socioeconomic phenomena rely on existing data and must take advantage of existing geographic units.

Census tracts, for example, can be defined as geographic units for a marketing analysis based on demographic characteristics. These boundaries are not appropriate for models of environmental and natural resource management because environmental conditions are usually not affected by administrative boundaries.

Delineation Based on Spatial Characteristics of the Data

Geographic units for environmental models are most adequately defined by the characteristics of natural units. GISs provide an effective tool for creating such units. For instance, in modeling wildfire distribution, neither administrative units nor an arbitrarily defined grid are suitable for defining geographic units. The most appropriate units are delineated by vegetation and topographic characteristics. For this reason, the vegetation and topography maps are overlaid to define polygons that are homogenous in both vegetation and topographic properties. These polygons represent natural units in fire distribution and are suitable for management purposes.

In general, the first step toward model building is to define geographic units. The analyst must carefully select the most appropriate method to delineate geographic units for the purpose at hand. The layer containing the geographic units becomes the base map for model building and spatial analysis.

Structural and Spatial Factors

Structural factors are variables that affect the "site" of each geographic unit. In other words, their measurements are confined to unit boundaries. For instance, when purchasing a house, structural factors that home buyers consider may include the size of the lot, the size of the house, the number of bedrooms, the age of the house, and so on. In modeling fire occurrence probability, structural factors include vegetation coverage, slope gradient, slope aspect, humidity, and so on.

Structural factors alone may not be sufficient to explain the distribution of certain spatial phenomena. In addition to variables that influence the "site" of each geographic unit, other variables that affect "situation" may be significant. Situation variables extend beyond the boundaries of the unit. In the house hunting example, home buyers may consider *spatial factors* such as public safety, scenic view, proximity to schools, and accessibility to freeways. For fire occurrence probability, spatial factors may include proximity to human structures, distance to roads, and neighborhood effects.

Spatial factors are measured in the following three ways: absolute location, relative location, and neighborhood effects. *Absolute location* is expressed in ground coordinates by longitude and latitude, or the x and y coordinates of any system. For instance, in a study of air transportation accessibility in the United States from 1960 to 1970 (Chou, 1993), the spatial pattern in the change of air transportation accessibility is shown to be unaffected by structural factors, such as income per capita or demographic growth. Instead, the spatial pattern is closely related to longitude, a spatial variable of absolute location.

Relative location of a geographic unit is evaluated by its distance from certain features. In modeling fire occurrence, for instance, proximity to roads and human structures is an important factor because human-caused fires tend to occur close to roads or human structures. Areas with identical environmental conditions may differ in fire occurrence probability solely because of locational variance relative to human structures. Another example of relative location is the significant correlation between the value of commercial properties and proximity to major attractions or public facilities. Because traffic volume tends to be higher for commercial establishments close to public transit stations, for example, rents are generally higher than for establishments at greater distances. In most spatial models, relative location plays a more important role than absolute location.

Neighborhood effects represent the influence of entities or features on similar entities in an adjacent area. The direction and extent of neighborhood effects are manifested in the spatial pattern. When attraction between entities plays a significant role, the existence of one entity tends to encourage other similar entities to locate nearby in order to take advantage of spatial agglomeration. Attraction among entities tends to form a clustered pattern in the distribution. An example is the common phenomenon of automobile dealership agglomeration. When repulsion plays a major role in distribution— the existence of an entity tends to discourage other similar entities from locating in adjacent areas—, the resulting spatial pattern tends to be scattered.

Neighborhood effects can be evaluated by spatial autocorrelation statistics. Among the available statistics, Moran's I coefficient (Moran 1948, 1950) is the most widely used. Formally, the I coefficient of spatial autocorrelation is defined as follows:

$$I = \frac{n\Sigma_i\Sigma_j\delta_{ij}(x_i - \bar{x})(x_j - \bar{x})}{S_o\Sigma(x_i - \bar{x})^2}$$

where $S_o = \Sigma_i\Sigma_j\delta_{ij}$, while n denotes the number of geographic units, and δ_{ij} represents the spatial relationship between units i and j in terms of adjacency. δ_{ij} equals 1 if the two units are adjacent to each other; otherwise it equals 0.

I coefficient values range from -1 to 1. When the existence of an entity tends to attract other entities, the distribution is described as *positive spatial autocorrelation*. In this case, the *I* coefficient will be in the positive domain, significantly different from 0. When entities tend to repulse each other, the distribution is described as *negative spatial autocorrelation*, and the *I* coefficient value will be negative. When neither attraction nor repulsion dominates the process, the spatial pattern is random, and the *I* coefficient is close to 0. In studies of wildfire distribution, fire activity has been shown to be clustered, and the *I* coefficient is significantly different from 0.

Because the variables are identified for inclusion in a model, their values must be translated to geographic units. Variables that are already measured based on geographic units can be directly copied to the attribute table for model building. For instance, if a proposed model is to be based on census tracts, then all census data variables can be directly copied to the attribute table. In many cases, GIS manipulation is required in order to obtain the data of variables that are not measured by the geographic units selected for the analysis.

Multiple Regression

Explaining the distribution of a spatial phenomenon requires the analysis of relationships between the phenomenon and potential explanatory variables. The two most useful methods in GIS spatial analysis for this purpose are multiple regression analysis and logistic regression analysis. Simple regression analysis aimed at examining the relationship between the phenomenon in question and a single explanatory variable is discussed in Chapter 5 and will not be treated here.

Logistic regression analysis is used to examine the relationship between the study phenomenon and multiple explanatory variables. This relationship is expressed as follows:

$$Y = \beta_0 + \beta_1 X_1 + \beta_2 X_2 + ... \beta_n X_n + e$$

where Y denotes the study (dependent) variable and Xs are explanatory (independent) variables; β_0 represents the intercept, while β_n denotes the estimated parameter of the variable X_n; and e is a randomly distributed error term.

For convenience, X_1 is equal to 1; thus, β_0 can be replaced by β_1. As such, the above equation can be substituted by the following matrix expression:

$$Y = X\beta + E$$

where Y is an nX1 vector with n equal to the number of observations; X is an nXm matrix with m equal to the number of independent variables (including the constant variable X_1); β is an mX1 vector; and E is an nX1 vector.

Minimization of the sum of squared errors yields the following expression:

$$\beta = (X'X)^{-1}X'Y$$

where X' is the transpose of the matrix X. Because multiple regression involves highly complex computation, in most spatial analyses a standard statistical package such as SAS, BMDP, or SPSS is used to estimate the parameters (βs).

The following example illustrates a typical environmental study of the relationship between bird distribution and selected environmental variables. The attribute table below lists a sample data set of 20 records. Each record is a polygon representing a geographic unit of the study. There are seven items in the table, including polygon ID and six variables labeled X1 through X6. The variables are defined as follows: X1 is a nominal code for denominating species; X2 denotes the frequency of observation (i.e., number of times birds are observed in the unit); X3 indicates the extent of forest coverage in the unit (forest density); X4 measures the distance from each unit to the nearest river (proximity to rivers); X5 represents the slope of the terrain in each unit; and X6 is a nominal variable of vegetation type. For convenience, the file containing the attribute table is named *Sample.dat.*

Sample.dat attribute table

ID	X1	X2	X3	X4	X5	X6
1	5	8	4	7	3	2
2	6	7	1	2	7	5
3	7	5	3	1	8	2
4	9	3	2	5	2	5
5	8	4	1	3	3	6
6	7	5	3	2	1	4
7	5	6	1	8	2	3
8	4	2	1	5	7	2
9	6	5	3	7	2	1
10	1	2	1	3	8	2
11	1	3	2	1	4	4
12	2	3	1	8	1	2
13	3	7	3	2	2	3
14	2	4	1	3	4	2
15	5	3	2	6	3	2
16	7	3	2	8	2	4
17	9	3	2	5	1	3
18	6	5	3	1	2	4
19	5	5	2	2	3	8
20	7	2	1	3	2	4

Because X1 and X6 are nominal variables, they are not considered in this study. A typical research question is whether the observational frequency of birds (X2) can be explained by forest density (X3), proximity to rivers (X4), and surface slope (X5). The data derived from the polygon attribute table can be analyzed using any statistical package such as SAS, BMDP, or SPSS. Appearing below is a simple SAS program used to construct a multiple regression model where the dependent variable is X2, and independent (explanatory) variables are X3, X4, and X5.

```
TITLE >Multiple Regression Sample Program=;
DATA TESTDATA;
INFILE Sample.dat;
INPUT ID, X1-X6;
RUN;
PROC REG;
        MODEL X2=X3 X4 X5;
RUN;
```

In the preceding block, the first line of the SAS program provides the program title. The second line specifies the name of the data set to be created in SAS. INFILE indicates that the input file is called *Sample.dat*. INPUT specifies that the data file contains seven attributes, ID and the six variables. RUN is used in SAS to execute the above command statements. PROC REG specifies that the regression procedure is to be carried out. Next, the MODEL statement specifies that X2 is the dependent variable and X3, X4, and X5 are the independent variables in the multiple regression model. Results of the above program appear in the next table.

Variable	DF	Parameter	Standard Error	t	Prob> |t|
intercept	1	2.0435	1.5495	1.319	.2058
X3	1	1.0469	0.4120	2.541	.0218
X4	1	-.0024	0.1623	-0.015	.9883
X5	1	0.0523	0.1814	0.288	.7771

According to the estimated parameters listed in the preceding table, the model can be expressed as follows:

$$X2 = 2.0435 + 1.0469\,X3 - 0.0024\,X4 + 0.0523\,X5$$

The model provides the estimated parameter for each variable; the estimated parameter can be used as the variable's relative weight in the distribution of the dependent variable. A key question is which variables incorporated in the model are significant. It is also important to determine which variables are not necessary. In general, an efficient spatial model is one that incorporates a small number of critical variables while generating a sufficiently accurate prediction. Significance tests can be performed to determine the structure of an efficient model according to the relative weight of each variable.

Significance Tests

The model's estimated parameters (βs) must be tested for statistical significance. In building spatial models, working with a database containing a large number of variables is not uncommon. Variables may be removed before model building if there are reasons to believe that they are not useful in the model. In addition, variables highly correlated with others are not necessary or useful. In order to determine the structure of the model, the *t*-test of statistical significance for estimated parameters must be performed.

The *t* quantity is formally defined as follows:

$$t = \frac{\beta}{S_\beta}$$

where S_β is the standard error of the estimated parameter computed by

$$S_\beta = \sqrt{\frac{S_e^2}{\Sigma(X_i - \bar{X})^2}}$$

The preceding table provides both the *t* statistics for the estimated parameters and the probability for the *t* distribution to have a value greater than the computed quantity. The *t* statistics are compared with the *t* table to determine the significance of the estimated parameters. In general, a .05 significance level can be selected from the *t* table for n - 2 degrees of freedom, where n is the number of records. In this example, n = 20; degrees of freedom are therefore 18. According to the *t* table provided in standard statistics textbooks, the critical value of the *t* statistic is 2.101 (α = .025 for d.f. = 18). Accordingly, only forest density (X3) is significant in the distribution of birds (X2). As such, both proximity to rivers (X4) and surface slope (X5) are discarded and the model can be reduced to the following expression:

$$X2 = 2.0435 + 1.0469 X3$$

To be more precise, a simple regression model with observational frequency (X2) as the dependent variable and forest density (X3) as the sole independent variable should be constructed again to obtain more appropriate estimates. The procedure described in this section shows how numerous variables in the original model can be evaluated for statistical significance. The result is a model which incorporates only significant variables.

Logistic Regression

On many occasions, the spatial phenomenon under investigation can only be described by a categorical variable. For instance, in the study of wildland fires, fire distribution is typically depicted by a map containing polygons that represent burned areas. The surface is divided into two categories, burned and unburned. Likewise, in the study of bird distributions, geographic units are labeled with a binary code indicating the presence or absence of the species under consideration. In such cases, the multiple regression technique discussed in the preceding section is not suitable because the dependent variable is neither interval nor ratio level. When the distribution (dependent variable) is measured at a nominal scale while explanatory (independent) variables are measured at an interval or ratio scale, the logistic regression method is appropriate for spatial modeling.

Conceptually, logistic regression treats the distribution in a probabilistic manner, that is, the occurrence of the study phenomenon is evaluated in terms of probability. If the probability of presence of the phenomenon is P_a, then P_b represents the absence of the phenomenon and $P_a + P_b = 1$. The logistic regression model is expressed as follows:

$$P_a = \frac{EXP(U_a)}{1 + EXP(U_a)}$$

$$U_a = \beta_0 + \beta_1 X_1 + \beta_2 X_2 + \ldots + \beta_n X_n + e$$

where U_a is a quantity commonly known as the utility function of event a, expressed as a linear combination of a number of explanatory variables, X_1, X_2, ... X_n; β_n is the estimated parameter of the variable X_n; and e is a randomly distributed error term.

In the logistic model, a greater value of U_a implies a greater probability for the event a to take place. The relationship between P_a and U_a is shown in the following diagram. When U_a approaches infinity, P_a approaches 1, indicating a high likelihood for the event to occur. When U_a approaches negative infinity, P_a approaches zero, and the chance for the event to occur is slim. When U_a equals zero, the probability is 0.50, implying a fifty-fifty chance for the event to occur. Thus, the probability of the event depends on the U_a quantity which, in turn, depends on the independent variables X_i for $i = 1, ...n$.

The S-curve represents the relationship between probability and utility in logistic regression models.

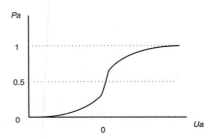

Most commercial statistical packages, such as BMDP and SAS, provide a logistic regression program. The program generates the estimated parameter of each variable along with a *p*-value for the testing of statistical significance. A smaller *p*-value implies a higher level of statistical significance, and a variable is considered significant at the .01 level if the *p*-value of its estimated parameter is less than .01.

For example, in a recent empirical study, fires that occurred between 1911 and 1984 in the San Jacinto Ranger District of the San Bernardino National Forest in California were examined in order to map the distribution of fire occurrence probability (Chou et al., 1990). Two models were compared, a basic model and a study model. The basic model consists of the eight independent variables listed in the next table.

Variables incorporated in fire distribution study

Variable	Definition
X_1	Area: area of the geographic unit.
X_2	Perimeter: perimeter of the geographic unit.
X_3	Vegetation: a vegetation computed by rotation period.
X_4	Building: proximity to structures.
X_5	Campground: proximity to campgrounds.
X_6	Road: proximity to roads.
X_7	Temperature: maximum temperature in July.
X_8	Precipitation: annual precipitation.

In this study, the dependent variable is a code indicating whether or not a geographic unit is burned or not. The probability of fire occurrence for each geographic unit is explained by the eight independent variables incorporated in the model. Area and perimeter of a geographic unit provide general geometric characteristics. Vegetation, temperature, and precipitation represent environmental factors that are meaningful to fire occurrence, while building, campground, and road represent human-related variables. The coefficients of these variables were estimated using the logistic regression program in BMDP. The results are shown in the following table.

Estimated coefficients for basic model

Variable	Coefficient	χ^2	P-value
X_0	-6.3246	31.13	0.0000
X_1	-0.0000	1.42	0.2340
X_2	-0.0002	8.13	0.0043
X_3	1.5577	43.65	0.0000
X_4	-1.1451	1.93	0.1648
X_5	-294.58	4.61	0.0318
X_6	-0.5244	4.46	0.0348
X_7	0.1790	28.19	0.0000
X_8	0.0023	0.21	0.6493

Log Likelihood = -1366.197

PCE = 60

$\chi^2 = 3.84$ for $\alpha = .05$ (d.f. = 1)

The results of the logistic regression include the estimated coefficients and associated χ^2 and p-values. The statistics are useful for determining the model structure and identifying the variables that are significant in the model. The estimated coefficient follows the χ^2 distribution. In other words, a small χ^2 implies a variable that is not significantly different from 0, and thus, the corresponding variable need not be included in the model. For a variable to be included in the model, its coefficient must be significantly different from 0; consequently, its χ^2 value must be at least greater than 95% of the χ^2 distribution, if the significance level is set at .05. The p-value provides the significance level. A p-value of .05 indicates that the coefficient is at the 95% level of the χ^2 distribution. In brief, for a variable to be accepted, its p-value must be less than or equal to .05. Accordingly, the model can be specified as follows:

$$U = -6.3246 \quad -0.0002 X_2 + 1.5577 X_3 - 294.58 X_5 - 0.5244 X_6 + 0.1790 X_7$$

The model indicates that perimeter, vegetation, campground, road, and temperature are the variables to be included in the model. Other variables are not included because they are not significantly different from 0. The statistical significance of a logistic regression model is examined through its level of log-likelihood.

However, a more convenient measurement of forecasting accuracy is the *percentage-correctly-estimated (PCE) index*. The PCE index shows the maximum level of estimation accuracy of a model. In this case, the PCE level indicates a highest possible level of forecasting accuracy of 60%. This level of accuracy is not satisfactory because a random guess of the outcome of burned or unburned is expected to generate an average accuracy of 50%. The model described above is defined as the basic model because it served as the basis for comparing the performance of other study models.

The alternative model incorporated an additional variable in order to determine whether it made any significant difference in model performance. This variable represents neighborhood effects, or conditions of the surrounding geographic units. The variable was added because fire occurrence probability is not only affected by the environmental and human-related variables listed in the basic model. Fire occurrence is also influenced by the distribution of fire occurrence probability of adjacent units. This spatial term, X9, is thereby defined by the percentage of the neighboring units that were burned during the study period. Consequently, this study model incorporates nine independent variables. The estimated coefficients and corresponding statistics are listed in the following table.

Estimated coefficients for study model

Variable	Coefficient	χ^2	P-value
X_0	-4.3602	1.36	0.2432
X_1	-0.0000	1.03	0.3106
X_2	-0.0003	0.97	0.3249
X_3	-1.6738	6.88	0.0087
X_4	-0.8416	0.19	0.6669
X_5	-42.280	0.00	0.9701
X_6	1.0241	3.00	0.0831
X_7	-0.1121	1.00	0.3168
X_8	-0.0127	0.55	0.4597
X_9	17.951	2359.3	0.0000
Log Likelihood = -164.788			
PCE = 97			
χ^2 = 3.84 for α = .05 (d.f. = 1)			

The results of the study model are quite different from those of the basic model. Only two variables are statistically significant, vegetation and neighborhood effects. The results indicate that vegetation is the determining environmental variable in the distribution of wildfires in the study area. In addition, the distribution of fire occurrence is significantly influenced by neighborhood conditions. The incorporation of the neighborhood variable improved the forecasting accuracy to a high level of 97%, indicating that the study model generates a high level of accuracy in explaining the distribution of fires.

The statistical significance of the study model compared to the basic model can be evaluated through the χ^2 test of likelihood ratio. In this case, the λ statistic is defined as follows:

$$\lambda = L_0/L_1 < 1$$

where L_0 denotes the likelihood of the basic model, and L_1 denotes the likelihood of the study model. This procedure is used to test whether the study model is significantly better than the basic model. The null hypothesis is that the study model is not significantly different from the basic model, which means that the incorporation of the additional variable does not improve the model significantly. If the latter is true, the additional variable is not necessary. However, the alternative hypothesis is that the study model likelihood is significantly higher than that of the basic model, implying a significant improvement of the model by adding the neighborhood variable. The log likelihood ratio is defined as Log λ = Log L_0 - Log L_1. The statistic -2Log λ has a χ^2 distribution.

In the above example, L_0 = -1366.197 and L_1 = -167.914. Thus, Log λ = -1198.283. The statistic -2 Log λ is equal to 2396.566, significant at the .01 level. Statistical testing suggests that incorporating the neighborhood variable significantly improved the performance of the fire occurrence probability model.

The same method was employed in a study of bird species distribution in Spain (Chou and Soret, 1996). The distributions of three bird species were examined, and logistic regression models were built to examine the role of neighborhood effects. The results suggested that neighborhood effects influence bird distributions differently according to species.

The next illustration shows the dominant species of mature ecosystems in Navarre, Spain, one of several data layers employed in modeling bird distributions. The area is classified into seven categories, including conifers, hydric oaks, beech, mesic oaks, riparian vegetation, Mediterranean oaks, and Mediterranean scrubland. Additional variables incorporated in the study included temperature, humidity, precipitation, hydrology, topography, cultivated land, and vegetation diversity, among others.

Dominant vegetation species of mature ecosystems in Navarre, Spain.

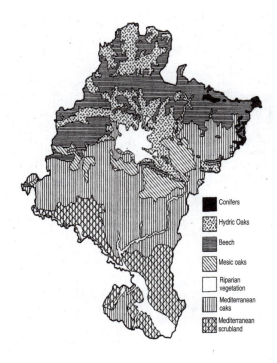

Conifers

Hydric Oaks

Beech

Mesic oaks

Riparian vegetation

Mediterranean oaks

Mediterranean scrubland

The following illustration shows the observed distributions of three bird species: European honey-buzzard, most in the northern central region; Eurasian hobby in the southern region; and European pied flycatcher in the northernmost corner.

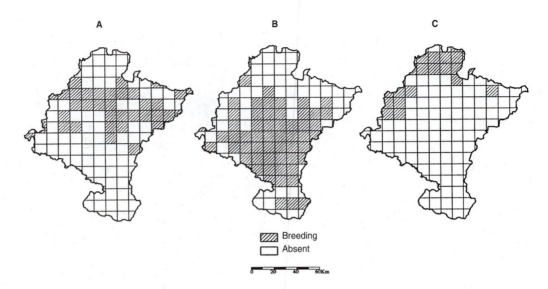

Breeding
Absent

Observed distributions of European honey-buzzard (A), Eurasian hobby (B), and European pied flycatcher (C).

The illustration below shows the distributions of the three bird species derived from logistic regression models. The spatial patterns clearly resemble those of observed patterns although minor differences are detected. In general, the spatial models generated satisfactory levels of forecasting accuracy: 87% for the European honey-buzzard, 82% for the Eurasian hobby, and 90% for the European pied flycatcher.

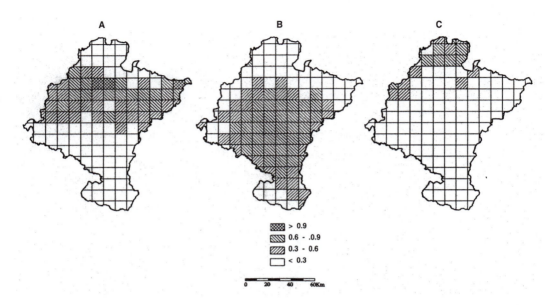

Model derived distributions of three bird species. The spatial patterns resemble those of the observed patterns depicted in the previous illustration.

Data Manipulation and INFO Programming

Manipulation of attribute data is usually executed in a database management system, and a GIS requires programming capability for database manipulation. The following examples show how attribute data can be manipulated or altered using INFO programs which interact with the spatial database in ESRI's ARC/INFO.

In ARC/INFO, polygon attribute data are organized in a database file called the polygon attribute table (PAT). Assume that a land information coverage is comprised of five polygons, resulting in five records in the following sample attribute table called *LAND.PAT.* Note that the original PAT always contains an additional record called the *universe poly-gon*. The universe polygon represents a hypothetical area defined by the outermost boundaries of the coverage, and its area is the negative of the total area of the coverage.

Next, a PAT file contains the following four basic items generated in ARC/INFO that characterize the geometric and topological properties of the polygons.

● An internal ID which is a unique code for each polygon assigned by the program

● A user-specified ID

● Polygon perimeter

● Polygon area

In addition to the above basic items, the user may define any number of polygon attributes. In the current hypothetical example, the basic items are not included for convenience. The file contains the following five attributes:

● POLYID, the polygon ID

● SLOPE, the average slope of the polygon

● VEGET, a vegetation variable

● PRECIP, a precipitation measure

● EROD, soil erodability (the study variable)

For the sake of convenience, units for the above variables are not discussed in the following examples.

The following table shows the initial PAT where the study variable is simply defined and no polygon values are assigned for this variable. Thus, the value of EROD is 0 for all polygons. Every polygon has known values for three independent variables: SLOPE, VEGET, and PRECIP. For instance, for polygon 1, SLOPE is 5.20, VEGET is 80.00, and PRECIP equals 10.10.

LAND.PAT

POLYID	SLOPE	VEGET	PRECIP	EROD
1	5.20	80.00	10.10	0.00
2	3.40	62.50	8.20	0.00
3	1.20	92.30	9.80	0.00
4	10.50	80.50	6.80	0.00
5	4.00	10.50	10.20	0.00

The following INFO program, *PROGRAM_1*, shows how to derive the value of EROD based on the following hypothetical model: EROD = 5.0 SLOPE - 0.5 VEGET + 0.2 PRECIP. The estimated parameters are obtained from either the multiple regression or logistic regression models described in previous sections of this chapter.

```
PROGRAM NAME: PROGRAM_1|  Program name specified for referencing.
PROGRAM SECTION ONE    |  Each program can be divided into sections.
SEL LAND.PAT           |  Select the file "LAND.PAT" for processing.
PROGRAM SECTION TWO    |  Start the second section.
CALC EROD = 5.0 * SLOPE - 0.5 * VEGET + 0.2 * PRECIP
```

```
                            | Calculate EROD according to the model.
PROGRAM END                 | End of INFO program.
```

As *PROGRAM_1* is compiled and executed, the EROD of each polygon in the PAT is assigned a value according to the specified model. Results are shown in the next table.

LAND.PAT for PROGRAM_1

POLYID	SLOPE	VEGET	PRECIP	EROD
1	5.20	80.00	10.10	-11.98
2	3.40	62.50	8.20	-12.61
3	1.20	92.30	9.80	-38.19
4	10.50	80.50	6.80	13.61
5	4.00	10.50	10.20	16.79

In spatial analysis, comparing spatial patterns based on models of different estimates is a frequent occurrence. For instance, if the estimated parameter of SLOPE in an alternative model is equal to 10.0 instead of 5.0, then the EROD for each polygon can be revised by executing *PROGRAM_2* below. The statements in this program are identical to those in *PROGRAM_1*.

```
PROGRAM NAME: PROGRAM_2
PROGRAM SECTION ONE
SEL EXAM.PAT
PROGRAM SECTION TWO
CALC EROD = 10.0 * SLOPE - 0.5 * VEGET + 0.2 * PRECIP
PROGRAM END
```

The next table shows the results of *PROGRAM_2*. The estimated values of EROD are different from those in the previous table.

LAND.PAT for PROGRAM_2

POLYID	SLOPE	VEGET	PRECIP	EROD
1	5.20	80.00	10.10	14.02
2	3.40	62.50	8.20	4.39
3	1.20	92.30	9.80	-32.19
4	10.50	80.50	6.80	66.11
5	4.00	10.50	10.20	36.79

It is not uncommon to derive a value based on specified conditions. For instance, if EROD is simply determined by the value of SLOPE, the following program illustrates how logical conditions can be incorporated using If statements.

```
PROGRAM NAME: PROGRAM_3
PROGRAM SECTION ONE
SEL EXAM.PAT
PROGRAM SECTION TWO
CALC EROD = 0.0          | Initialize the value of EROD.
IF SLOPE GE 10.0         | If SLOPE is greater than or equal to 10.0,
      CALC EROD = 9.9    | EROD is assigned the value 9.9.
ELSE                     | Otherwise, EROD is assigned a different value.
IF SLOPE GE 5.0
      CALC EROD = 7.0
ELSE
IF SLOPE LT 5.0
      CALC EROD = 0.0
ENDIF                    | Each IF statement must be closed with an ENDIF.
ENDIF
```

```
ENDIF                   | With three IFs, there must be three ENDIFs.
PROGRAM END
```

The next table shows the results of *PROGRAM_3*. As expected, the estimated values of EROD are determined by the specified conditions of SLOPE.

LAND.PAT for PROGRAM_3

POLYID	SLOPE	VEGET	PRECIP	EROD
1	5.20	80.00	10.10	7.00
2	3.40	62.50	8.20	0.00
3	1.20	92.30	9.80	0.00
4	10.50	80.50	6.80	9.90
5	4.00	10.50	10.20	0.00

Sometimes logical conditions are more complicated than can be expressed in a single statement. In this case, the most effective way to specify conditions is by grouping statements of complicated conditions in a set and executing the set of statements whenever conditions satisfy the criteria. The set of statements can be labeled and executed using the GOTO statement.

```
PROGRAM NAME: PROGRAM_4
PROGRAM SECTION ONE
SEL EXAM.PAT
PROGRAM SECTION TWO
CALC EROD = 0.0
IF SLOPE GE 10.0        | If SLOPE is greater than or equal to 10.0
        GOTO HIGH       | Go to the set labeled "HIGH."
ELSE                    | Otherwise,
IF SLOPE GE 5.0         | if SLOPE is greater than or equal to 5.0,
```

```
        GOTO MED           | Go to the set labeled "MED."
ELSE
            GOTO LOW
ENDIF
ENDIF
LABEL HIGH                 | The set labeled "HIGH."
      CALC EROD = 9.9      | Calculate EROD.
      GOTO PROGEND         | Move on to end of program.
LABEL MED
      CALC EROD = 7.0
      GOTO PROGEND
LABEL LOW
      CALC EROD = 0.0
LABEL PROGEND
PROGRAM END
```

The results of *PROGRAM_4* are identical to those of *PROGRAM_3* because their logical conditions are identical.

Relational tables are also commonly employed in INFO programming. The following look-up table, *SLOPE.LUT,* relates the values of SLOPE to another measure called EFACT. Look-up tables are primarily used to translate measurements. For instance, slope gradient can be translated into another measurement more suitable for evaluating soil erodability. If SLOPE is greater than or equal to 20.00, then the measurement related to soil erodability (EFACT) is

5.00. Otherwise, if SLOPE is greater than or equal to 10.50, EFACT equals 3.00.

SLOPE.LUT

SLOPE	EFACT
20.00	5.00
10.50	3.00
5.20	2.00
4.00	1.50
3.40	1.00
1.20	0.40
0.00	0.00

With the *SLOPE.LUT* look-up table, the EROD variable can be computed based on the value of EFACT. *PROGRAM_5* demonstrates how this is accomplished.

```
PROGRAM NAME: PROGRAM_5
PROGRAM SECTION ONE
SEL EXAM.PAT
REL SLOPE.LUT 1 BY SLOPE | Define Relation 1 to SLOPE.LUT
                         | by the common item "SLOPE."
PROGRAM SECTION TWO
CALC EROD = $1EFACT      | EROD is derived from Relation 1 through
                         | the value of "EFACT."
PROGRAM END
```

The table below shows the results of *PROGRAM_5*.

LAND.PAT for PROGRAM_5

POLYID	SLOPE	VEGET	PRECIP	EROD
1	5.20	80.00	10.10	2.00
2	3.40	62.50	8.20	1.00
3	1.20	92.30	9.80	040
4	10.50	80.50	6.80	3.00
5	4.00	10.50	10.20	1.50

Another useful feature in INFO programming is the ability to run one program from another. *PROGRAM_6* shows how *PROGRAM_1* can be called and executed.

```
PROGRAM NAME: PROGRAM_6
PROGRAM SECTION ONE
SEL EXAM.PAT
RUN PROGRAM_1 LINK        | Execute PROGRAM_1 and return (LINK) the result.
PROGRAM SECTION TWO
CALC EROD = EROD + 10.0   | Further modify the EROD value.
PROGRAM END
```

The results of *PROGRAM_6* appear in the following table.

LAND.PAT for PROGRAM_6

POLYID	SLOPE	VEGET	PRECIP	EROD
1	5.20	80.00	10.10	-1.98
2	3.40	62.50	8.20	-2.61
3	1.20	92.30	9.80	-28.19
4	10.50	80.50	6.80	23.61
5	4.00	10.50	10.20	26.79

The preceding sample programs illustrate how the database in a GIS can be manipulated using the programming tools in the database management system. Such tools are especially useful in spatial modeling when spatial patterns based on parameters generated from different models are compared and evaluated.

Summary

One of the most important functions of GISs is to answer "what if" questions about spatial phenomena. For instance, the question "what will happen to the natural habitat of an endangered species if a natural forest is developed into a recreational park," can be effectively answered with a spatial model incorporating the distribution of the species, environmental conditions pertaining to the habitat of the species, the forested area, and the planned park. In this regard, spatial models that deal with such questions can be constructed for several purposes.

First, spatial models are used to describe the spatial pattern in the distribution of a phenomenon under investigation. Second, they explain relationships between the distribution of a phenomenon and variables that are significantly correlated with the distribution. Third, once explanatory models are built and parameters estimated, predictive models can be used to estimate changes in the distribution in response to the alteration in any significant variable. Fourth, if necessary, normative models can be formulated for developing the most cost-effective spatial strategies for management of system planning.

In building spatial models, key issues include delineation of geographic units, specification of both the dependent variable and explanatory variables, formulation of the model, testing the statistical significance of the model, differentiation of significant variables from insignificant ones, and interpretation of results. "What if" questions can then be analyzed by altering the environmental variables associated with the proposed "if" conditions. In general, multiple regression and logistic regression are most commonly employed in spatial modeling. When the dependent variable and explanatory variables are measured at an interval or ratio scale, the multiple regression technique is adequate. The logistic regression technique is used when the dependent variable is measured at a nominal scale. In both cases, the estimated parameters must be tested for statistical significance in order to identify variables that are significant in the distribution of the study phenomenon.

When using a GIS to build spatial models, data organized in attribute tables must be manipulated and converted throughout the modeling process. The programming tools in a GIS database management system are especially useful in preparing data for analysis. INFO programs are required for data manipulation tasks in ESRI's ARC/INFO. This chapter provided simple programs to illustrate how data can be manipulated in the modeling process.

Exercise

1. Digitize the five maps depicted in the diagram below. Map A shows the boundaries of demo-

graphic statistical zones (e.g., census tracts). Map B shows the distribution of crime incidents per unit area. Map C shows the density of housing, measured by the number of dwelling units per unit area. Map D shows the distribution of average property values per unit area. Map E shows point locations where major commercial establishments and public facilities are located.

A

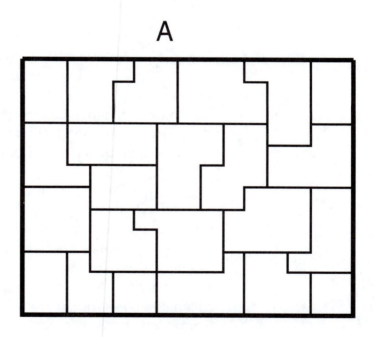

B

C

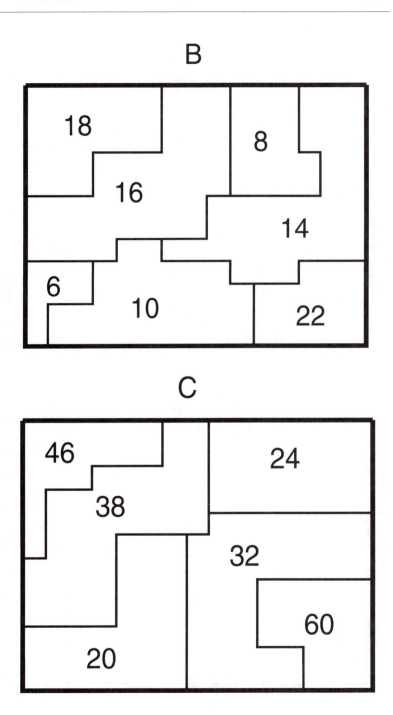

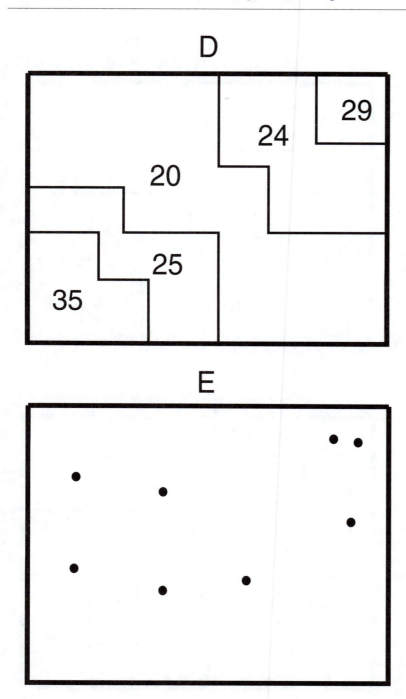

2. Overlap maps B, C, and D on map A. Obtain the crime rate, housing density, and average property value for each demographic statistical zone using the IDENTITY procedure described in Chapter 5.

3. Calculate the proximity to commercial establishments or public facilities for each zone using the NEAR function described in Chapter 5. This quantity is the distance from the centroid of each zone to its nearest point location (a commercial establishment or public facility).

4. Use INFO programs to manipulate the polygon attribute table (PAT) of Map A in order to obtain the value of each variable for each demographic statistical zone.

5. Build a multiple regression model with crime rate as the dependent variable, and housing density, property value, and proximity to commercial establishments or public facilities as independent variables. Use a statistical package to estimate the parameters and test the statistical significance of each variable.

6. Use INFO programs to generate a map showing the distribution of crime rates based on the model-generated parameters. Compare results of the model with the empirical distribution.

Surface Analysis

Surface analysis entails analyzing the distribution of a variable which can be represented as the third dimension of spatial data. Therefore, in addition to the required locational information about the x and y coordinates of the area, surface analysis involves a third quantity which represents the variation of the surface recorded on the Cartesian plane. *Elevation* is a natural example of the third-dimensional variable because variation in elevation produces the realized surface of the Earth.

Conventionally, the 3D attribute is called the *z variable*. The z variable can be anything from a physical attribute to socioeconomic variable, as long as the attribute is quantifiable. Physical attributes such as precipitation, elevation, and temperature, and socioeconomic variables such as income, crime rate, and land value, are possible z variable candidates. For ease of explanation, elevation data representing terrain topography are used as examples in this chapter.

This chapter presents two major issues pertaining to surface analysis. First, information about a surface can be organized in a GIS in several ways. How such information is organized and processed is presented in the first section. Second, the underlying assumption in surface analysis is that a surface is continuous, while the representation of the surface in digital data is always discrete in nature. The most fundamental processing of surface information is *spatial interpolation*. Spatial interpolation allows a continuous surface to be constructed from a set of discrete data points and is the main subject of the second section. The third section describes common applications of GIS-based surface analysis.

Organization of Surface Information

In vector-based GIS, z variable data—which characterize a surface—are organized in one of the following ways, depending on the nature of the data source.

- The surface may be represented by a set of irregularly-spaced point features.

- A simple, raster-based data structure may be used to record the surface by a set of regularly spaced data points that form a grid, also known as a *lattice*.

- The surface may be represented by a set of digitized contours as line features.

- Surface information may be organized into a set of triangulated irregular networks (TINs).

In addition to the above forms, z variable data may also be organized as polygon features, although this is not practical, and thus not usually employed in GISs.

Irregularly Spaced Points

In its simplest form, surface information can be organized in a set of irregularly located points. In this case, each record represents a point on the map. The point's x and y coordinates specify its location, and the z variable specifies its volumetric attribute. In terms of data storage, this is the most efficient organization of surface information because data volume can be reduced by incorporating only a minimum set of data points. Any vast, flat surface or constant slope area can be efficiently represented by a few point features delineating the outline of the area. This type of data organization is commonly used by land surveyors. Topographic surveys can hardly be performed at regularly spaced points due to terrain constraints, accessibility, and economic considerations. For large areas of similar topography, however, it is not necessary to survey the entire elevation because the entire surface can be efficiently represented by a smaller number of data points.

Although organizing the z variable by irregular data points is cost-effective, the tradeoff is that the data always require processing before they can be used for either presentation or analysis. A map showing point features labeled with elevation does not show the spatial distribution of the elevation, creating presentation problems. Constructing the surface directly from a set of irregularly spaced points is very difficult; therefore, this step must be completed before analysis. The following illustration shows a typical map of irregularly spaced points of surface information.

Surface information can be organized by irregularly distributed points.

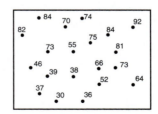

Tessellation

Effective processing of surface information in a GIS requires the translation of any surface into smaller areas of relatively homogeneous characteristics. The organization of the surface into a set of organized polygons is called *tessellation*. Tessellation can be accomplished in different ways. The next illustration shows three common cases of tessellation based on hexagons, squares, and triangles. In these cases, the unit area is the same for each figure type (i.e., in the hexagonal tessellation, all hexagons have equal area).

Three typical configurations of surface tessellation.

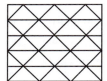

The hexagonal tessellation is the natural structure found in bee hives. An important property of this structure is that every hexagon is of equal distance from the six adjacent hexagons, a property not available in any other tessellation. However, the hexagons are intertwined, and thus centroid locations of hexagons are aligned in only one dimension, which is vertical in the example. A major weakness of hexagonal tessellation is that hierarchical tessellation is impossible without creating new hexagon systems. In other words, combining multiple units of hexagons to form a larger hexagon is impossible. This weakness reduces the usefulness of this form for processing surface data.

The squared tessellation has centroid locations of units aligned both vertically and horizontally. The square system is identical to a grid, and is widely employed by GIS users because of two important properties. First, all orthogonal units are perfectly aligned, and thus, they can be referenced systematically by row and column counts. Second, each group of four units forms a larger unit, while the system of the larger units bears the identical data structure as that of the smaller units. In other words, hierarchical tessellation is a natural process in the square system.

A minor problem with the squared system is that for each square unit, the adjacent units are of two types: orthogonal and diagonal. The distance between orthogonal neighbors is different from the distance between diagonal neighbors. Regarding the encoding of spatial relationships between adjacent units, the squared tessellation is more complicated than either hexagonal or triangular tessellation because of the need to differentiate between the two types of neighbors. Another issue affecting the utility of the square tessellation is known as the *saddle point* problem, that is, the centroid location may have two possible estimates based on the values of the four corners.

In triangular tessellation, the centroid locations of adjacent units are aligned vertically, but not horizontally. An important property of triangular units is that the shape is geometrically indivisible. In brief, a triangle cannot be divided into subunits without creating new vertices. This property ensures that there is no saddle point problem with the triangle configuration. In addition, hierarchical tessellation is possible with triangles.

The main purpose of tessellation in GIS is to delineate area features for surface analysis. With the tessellation capability, point data can be converted into polygons, and the surface information is then organized into the polygon features. The above systems of tessellation, however, are limited to regularly spaced data points. In many cases, data points are not regularly spaced, and a different type of tessellation must be implemented.

Surface tessellation allows for the z variable to be recorded in a systematic way where data points are regularly spaced for easy referencing. For instance, if the triangle configuration is adopted, elevation data may be recorded for every triangle vertex. Among the three possible configurations, the rectangle/ square tessellation is most commonly used because the shape can be easily constructed and the location of every square can be clearly referenced.

The rectangle/square configuration is the simplest and most common method of surface tessellation, wherein the z variable is arranged in a gridded, raster format. The most commonly used digital elevation data, the U.S. Geological Survey's 7.5-minute digital elevation models (DEMs), are organized in a gridded format at a constant spatial separation of 30 m. The gridded format has the advantage of easy encoding and effective referencing. Because they are all regularly spaced, the x and y coordinates are systematically coded and can be easily translated into columns and rows in a data set. Consequently, referencing x and y coordinates can be drastically simplified. A related advantage is that the entire surface is evenly covered by data points, and thus random sampling can be performed directly from the database. The following table shows the DEM for a portion of the El Cajon quadrangle in California.

Sample data of 7.5-minute DEM, El Cajon

1183	1186	1190	1193	1180	1167	1153	1140	1137	1134	1131	1129	1131	1135	1139
1143	1146	1149	1151	1153	1146	1137	1129	1121	1107	1093	1079	1065	1058	1051
1044	1038	1037	1037	1037	1037	1036	1035	1034	1034	1035	1037	1039	1041	1043
1044	1045	1046	1050	1054	1058	1061	1061	1060	1060	1060	1066	1071	1076	1081
1081	1082	1082	1083	1089	1095	1101	1107	1119	1130	1142	1153	1164	1175	1186
1196	1196	1195	1195	1194	1192	1189	1187	1185	1187	1190	1192	1193	1188	1184
1170	1172	1175	1177	1166	1153	1140	1127	1123	1120	1118	1115	1122	1130	1139
1147	1147	1145	1143	1141	1135	1129	1123	1117	1103	1089	1076	1062	1055	1049
1043	1037	1037	1036	1036	1035	1034	1033	1031	1030	1032	1034	1037	1039	1038
1038	1038	1038	1041	1044	1047	1049	1051	1052	1053	1054	1058	1062	1066	1070
1071	1071	1072	1072	1079	1085	1091	1097	1108	1118	1129	1139	1150	1161	1172
1182	1181	1180	1179	1178	1177	1177	1177	1176	1177	1178	1179	1180	1176	1171
1153	1155	1158	1161	1150	1137	1125	1113	1109	1106	1104	1102	1112	1124	1136
1149	1146	1141	1135	1129	1125	1121	1117	1114	1100	1085	1069	1054	1048	1044
1040	1036	1036	1037	1038	1038	1036	1034	1031	1028	1029	1030	1030	1031	1033
1034	1036	1038	1039	1040	1041	1043	1045	1047	1049	1051	1054	1057	1060	1063
1063	1064	1064	1064	1070	1077	1084	1091	1100	1109	1117	1126	1137	1147	1158
1168	1166	1164	1161	1159	1159	1160	1161	1162	1163	1164	1164	1165	1161	1158
1127	1131	1135	1139	1132	1121	1110	1100	1096	1093	1090	1087	1100	1115	1131

Digitized Contours

A third method for organizing the 3D attribute is digitized contour lines. In this case, the z variable data are converted into line features of the same value. A contour is a line of equal value to the z variable. Essentially, the surface is represented by a set of contour lines of different values at a constant interval. The U.S. Geological Survey's digital line graph (DLG) topography data are organized in this format.

Data organized in the above format have the advantage of being readily presentable, in that contour lines provide an effective means for showing the distribution of the z variable. Thus, contour plots are a common presentation method. Another advantage of the digitized contour method is efficiency of data storage (Chou, 1992). However, a major disadvantage is that raw data must be processed in order to reorganize data into the digitized contour format. In addition, conducting GIS analysis on digitized contours is generally more difficult than on raster format data.

The following figures show elevation point feature and contours derived from the same data. The point locations are shown as dots for referencing purposes only. The data file needs to contain four records, each representing a line segment. In this example, there are two lines of the value 80, one line of 60, and one of 40. Each record contains a sequence of x and y coordinates of the vertices that delineate the line, and a z value such as 80, 60, or 40.

*Elevation point
feature data.*

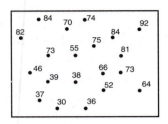

*Surface information can
be effectively organized
in digitized contours.*

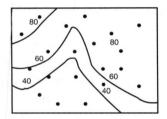

Triangulated Irregular Networks (TINs)

The z variable can also be organized in the form of triangulated irregular networks (TINs). In this case, the surface is converted into a set of triangular facets and the data are stored in the form of polygons that have degenerated into triangles. Triangular facets are of the most basic geometric shape in the sense that a triangle cannot be divided into subdivisions without creating additional nodes. The z variable can be organized at the vertices of the triangular facets such that both the slope and aspect (slope orientation) of the facet are computable.

The TIN structure incorporates the advantage of irregularly spaced points in that the density of the points is dependent on terrain. In other words, areas of great variation in terrain require a higher density of data points to more accurately represent the topography, while areas of simple terrain can be more efficiently represented by fewer data points. An obvious disadvantage is that the structure requires more data elements and becomes uneconomic for small areas. Another disadvantage is that TINs can only be generated to cover an area within the outer bound of the data points, the so-called *convex hull*. In most applications, the outer bound needs to be expanded to ensure complete coverage of the convex hull for the study area.

The next figure shows the TIN generated from the previous point coverage. Surface information is encoded in the triangular facets, and elevation data at the vertices of each triangle.

Triangulated irregular network (TIN) generated from the previous point coverage.

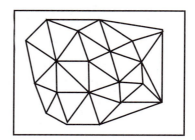

Spatial Interpolation

Spatial interpolation estimates of the z variable value for any point location within the map area based on the available data of known locations. Two fundamental assumptions underlie spatial interpolation. First, the surface of the z variable is continuous. Therefore, the data value at any location can be estimated if sufficient information about the surface is provided. This assumption is explicit and allows for spatial interpolation methods to be formulated. Second, an implicit assumption is that the z variable is spatially dependent. In brief, the interpolation of the variable value can be extracted from the given spatial distribution because the value at any specific location is related to the values of surrounding locations. The second assumption is important in the reasoning of estimation methods. For instance, the simplest expression of this assumption is spatial autocorrelation, such that the z variable at the specific location is affected more by nearby locations than by distant locations.

Interpolation from Contours

The following illustration depicts an example of spatial interpolation on a contour map. Contours are labeled with the z variable at a constant interval of 20 m. Point A is located somewhere between the 120 m and 140 m contour lines. Because the surface is assumed to be continuous, and the change at the locality is generally constant (on a constant slope according to the pattern of contour lines), the value at point A can be estimated based on the ratio of the distance between point A and the 120 m line to the distance between point A and the 140 m line. If point A is closer to 120 m than to 140 m, the estimated value should be less than 130 m. If point A is located exactly halfway between these two contours, the estimated value is 130 m.

Spatial interpolation based on the shortest segment method.

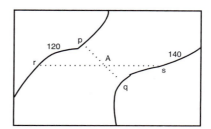

Distance is measured from the shortest possible segment connecting the point location and the two adjacent contours. In this example, two line segments, p-q and r-s, are considered. Both segments connect point A to the adjacent contours. Because the p-q segment is shorter than r-s, the distance should be measured on the p-q segment. In other words, the interpolation of surface value at any location on a contour map requires the construction of the shortest straight line which connects the specified location to the adjacent contours. Once such a segment is found, the segment can be divided into sections equal to the contour interval, and the proportional distance from the location to either contour can be estimated.

Gridding

Z variable data, when organized in irregularly spaced points, are not appropriate for graphical presentation nor are they suitable for analysis. Further processing for surface analysis requires that z variable data be converted into either a regularly spaced system or a set of triangulated irregular networks. Gridding is the procedure used to convert irregularly spaced data to a regularly spaced format, which then serves as an intermediate format for the organization of the z variable. Spatial interpolation based on TINs will be discussed in a subsequent section.

Local Estimation

Gridding involves the interpolation of the z value for any point of known location, usually at the intersection of vertical and horizontal lines, from a set of irregularly spaced data points. The two general approaches for gridding are local estimation and global approximation.

Local estimation means that the z value of a specific point location is estimated from the data of its neighborhood. In other words, only the known values of the adjacent area are used in estimating the value at this location. An obvious advantage of this approach is that the estimation is based on surrounding information and more accurately reflects variation in the immediate neighborhood. A disadvantage is that a priori knowledge about the suitable scope of the "valid neighborhood" must be provided. Next, methods for deciding how many points of known data, and which points should be used for estimation, should be selected. Furthermore, surface variation at a particular locality may be affected to some degree by data values in distant locations.

The alternative approach is *global approximation*, which estimates the z value at any location based on the entire set of known data. The advantage to this approach is that the estimation procedure applies to the entire map area. Consequently, the quality of the estimate depends only on the global estimation methods employed. In other words, the estimation is not affected by the number of points and the selection of such points in the estimation procedure.

For local estimation, the first step is to decide on the number of points to be used for estimation. In this scenario, some knowledge about the available data and the study area is required. For instance, if the data are relatively dense (i.e., numerous data points are available everywhere), there is no need to pick a large number of data points for estimation. Next, if the study area has a relatively uniform distribution of the z value (e.g., a flat surface), then the number of selected points can be small. On the other hand, if the density of existing data points is low, the number of points may have to be increased for a more accurate estimation.

Once the number of points is identified, the next question is how to select the points from existing data. *Local search* is the general procedure for making such decisions. Local search methods include nearest neighbor, quadrant, octant, and various radius. The following illustrations show examples of the four search methods. In each illustration, there are 15 irregularly spaced data points. The center in each block, labeled with a cross hair, is the location where the z value is to be estimated from the surrounding data points. In these examples, eight data points will be used for estimating the z value.

Nearest neighbor is the most commonly used method. In this method, the points closest to the point location for estimation are selected. Because eight points will be used for estimation, the analyst must find the eight points closest to the cross hair location, regardless of how these points are distributed. In the illustration, the selected points are indicated by squares, while those not selected are labeled with diamonds.

Nearest neighbor search method.

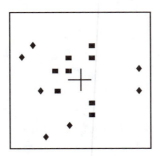

The nearest neighbor method requires a given measure of distance. In most applications this method is adequate; however, when the surface is systematically patterned, the selected points may not be the best possible set for estimation. For instance, if many near points fall in the same direction from the point of estimation, the selected set may be biased due to the fact that variations in other directions are neglected.

The quadrant search is based on a system in which the entire map area is divided into four quadrants centered at the estimation point. The illustration shows an example in which eight points are to be selected for estimation. An equal number of points are selected from each quadrant. With eight points to select, two must be selected from each quadrant. Within each quadrant, the selection criterion is nearest neighbor. This method avoids the common problem of most selected points coming from the same direction, which introduces directional bias in the estimation procedure. On the other hand, a disadvantage is that some points at a much greater distance from the point of estimation may be included.

Quadrant search method

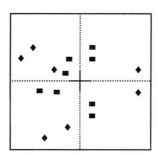

The octant search is an extension of the quadrant search. The locality surrounding the point of estimation is divided into eight partitions, and then an equal number of points are selected from each partition. In general, when the input points are dense enough and evenly cover the area, this method tends to produce better results than either the quadrant search or the nearest neighbor method. On the other hand, for locations near the edge of the map area, it is difficult for this method to obtain enough data points from all the eight directions.

Octant search method.

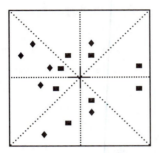

In the various radius method, a certain radius from the point for estimation is pre-specified. The search is limited to the area within the specified radius. If more than the required number of points are in the search range, then either all points within the specified radius are selected or the nearest neighbor method is used to select a precise number of points. If the number within the search range is less than the required number, then a specific value will be given to extend the radius until the required number of points are found. In the illustration, the first radius identifies only one point for selection. Once the radius is enlarged twice, a total of nine points fall inside the search radius. In the example, all nine points are selected.

Various radius search method.

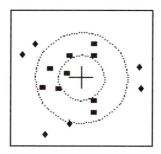

Estimation

Once the required number of data points is found, the next step is to estimate the z value of the point based on the values of the selected points. The estimated value may vary significantly depending on the estimation method. Therefore, the analyst must be knowledgeable about the estimation method employed in the GIS in order to have a good sense of the quality of the estimated surface.

The simplest estimation is the *unweighted mean*, that is, the estimated value is the average of the data points selected. This method is straightforward, but is not accurate for most applications. The unweighted mean method is based on the *isotropic* assumption that no directional trend exists in the surface. In other words, if all data points are spatially independent, the average of selected points could be a useful indicator of the estimated value. If, for instance, the data on one side of the point to be estimated are all higher than the data on the other side of the point, then the average is a valid estimate of the point.

In other situations the isotropic assumption may not be valid. For instance, if all points show a tendency that closer points are at lower elevations, while farther points are higher, then the surface trend appears to indicate a depression in the center. In this example, the average will give a value which is quite different from the surface trend.

The above problem can be partially offset by using the *distance weighted method*. This method dictates that the closer points are assigned a greater weight than farther locations in estimating the z value. A key question for this method is specification of the distance function. Distance functions assign weights in different ways. For instance, the relationship between spatial weight and distance can be specified by a squared inverse or a cubic inverse function. Other functions are available, and there are no simple selection criteria. The function most suitable for one surface may not be the best for another.

A more advanced and complicated method is to estimate the value based on both the average slope and the z value of each selected data point. This method requires a two-phase estimation procedure. In the first phase, the average slope at each data point is derived from its neighboring points. The average slope is called the *local dip* of the data point. The estimated value at a grid node is thus the average of the projected values of the selected points. The next figure illustrates this procedure.

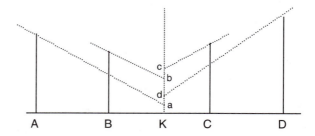

In a two-phase projection procedure, the estimated value at location K is the average of projected values.

In the preceding figure, the z value at point K is to be estimated from four points (A, B, C, and D). The two-phase projection procedure requires that the slope at each of the known locations be generated first. The slope at point A is projected to the K location at the *a* value. Likewise, the projected value of point B is *b* at location K. The second phase of the procedure is to take the average of the four projected values as the estimated value of K.

Triangulation

The second approach for converting irregularly spaced data points into a format suitable for spatial analysis is through triangulation. In general, triangulation means the conversion of points into a set of triangular facets that completely cover the map area. Each triangular facet is assumed to be homogeneous in spatial properties. In other words, the z value on a triangular facet varies from place to place at a constant gradient.

A set of data points can be triangulated (via the "greedy" triangulation method) by connecting every point with its closest adjacent points by non-crossing, straight line segments. Once all possible connections—without crossing existing segments—are built, the system is a set of triangular facets of irregular size. Because data values are available for all vertices of each triangle, the surface of the triangle can be characterized by slope gradient and orientation (aspect).

Triangulation can be carried out in different ways depending on the criterion for connecting vertices. The following illustration shows an example of vertex connections using the Delauney method. When the four vertices (A, B, C, and D) are to be connected, there are two possible configurations. *Maximum-minimum angle* is a commonly adopted criterion to determine which configuration to follow. Each configuration generates six internal angles inside the two triangles. The minimum angle of the left side configuration is the angle specified by the D-B-A vertices, while the minimum angle on the right side configuration is D-A-C. Because the D-A-C angle is larger than D-B-A, the configuration on the right side is adopted. A similar method is based on the shortest segment criterion. In this case, the connection of all vertices is based on the shortest possible segments. Because the AC segment is shorter than the DB segment, the right side configuration is used.

Each set of four points may derive two possible triangulations.

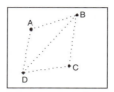 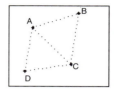

Spatial interpolation on a TIN is based on either a simple linear function or a more complicated bivariate quintic function. The simple linear function assumes that the surface of a triangular facet is of homogeneous, constant slope. In this case, the z value at any location within a triangular facet is a function of its x and y coordinates, such that

$$Z = aX + bY + c$$

Interpolation based on the linear function is simpler, and the surface is similar to the triangular facet at the left of the next illustration. In this case, the surface has a constant slope between any pair of vertices. The bivariate quintic interpolation assumes that a complicated surface can be represented by a polynomial equation to the fifth power of both x and y coordinates, such that

$$Z = \beta_1 + \beta_2 X^5 + \beta_3 X^4 Y + \beta_4 X^4 + \beta_5 X^3 Y^2 + \beta_6 X^3 + \beta_7 X^3 + \beta_8 X^2 Y^3 +$$

$$\beta_9 X^2 Y^2 + \beta_{10} X^2 Y + \beta_{11} X^2 + \beta_{12} XY^4 + \beta_{13} XY^3 + \beta_{14} XY^2 + \beta_{15} XY$$

$$+ \beta_{16} X + \beta_{17} Y^5 + \beta_{18} Y^4 + \beta_{19} Y^3 + \beta_{20} Y^2 + \beta_{21} Y$$

In the bivariate quintic function, there are 21 parameters to be estimated, and the surface interpolated from this function can be of greater variation than the simple function. The diagram at the right in the following illustration shows a typical surface interpolated using the bivariate quintic function. The slope between two vertices is a complicated curve depending on the parameters estimated from the regression.

Two types of surface interpolation.

The following diagram shows triangulated irregular networks generated from the digital elevation models of the San Jacinto Ranger District in California. Each triangular facet represents a surface of homogeneous topographic characteristics.

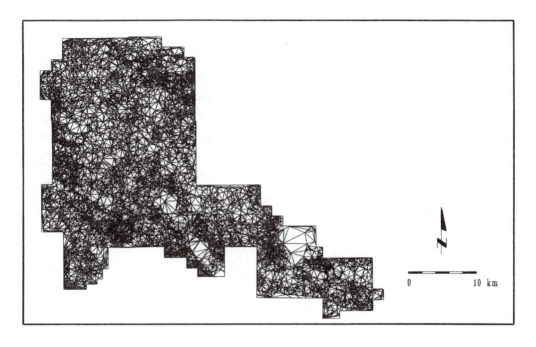

Triangulated irregular networks derived from digital elevation models of San Jacinto Ranger District, California.

Global Approximation

Global approximation implies that the estimation of a data point is based on the entire array of available data. There are two general approaches to global approximation: the polynomial trend surface method and kriging.

Polynomial Trend Surface Method

The polynomial trend surface method, similar to estimation in triangulation, requires the estimation of the parameters of a predetermined polynomial function using all available data. The number of parameters determines the degree of detail the generated surface may exhibit. For instance, the simplest form is the first order function in which there are only three parameters to estimate, identical to the simple linear function in the triangulation. In this case, the surface is assumed to be constantly dipping, and the three parameters to be estimated represent the characteristics of this surface. Each additional parameter introduces another possible reflection on the surface.

The most widely used method is Akima's bivariate quintic function which requires 21 parameters to be estimated for a function up to the fifth power for both the x and y values. The number of parameters to be estimated for any surface, predetermined by the analyst, dictates the characteristics of the surface.

Kriging

In geostatistics terms, kriging is the optimal method of spatial linear interpolation, where the mean is estimated from the best linear unbiased estimator or best linear weighted moving average. In theory, kriging generates estimates of surface values that are the most accurate among available methods of spatial interpolation. However, the effectiveness of kriging depends on the selection of parameters that determine the behavior of the covariance between data points. Kriging involves much more complex computations than any other interpolation method.

The estimation of surface variations is based on the assumption that the surface can be represented by two factors: the drift of regional tendency and the residual of local fluctuation. The regional trend can be estimated from a polynomial function such as the bivariate quintic function described earlier. Theoretically, if the regional trend is removed, the surface is considered stationary and the only variable left is the residual. The relationships between residuals are then represented by a semivariogram which relates each data point to all other data points with respect to the distance between them.

The semivariogram can be expressed as follows:

$$\gamma_d = \omega d$$

where γ_d represents the semivariance (one half of the variance) over distance d. The function is assumed to be identical for all directions.

Prior to the application of kriging, the data must be carefully examined and the slope of the semivariogram determined. The order of the polynomial function representing the drift must also be specified. The 0-order drift can be expressed as $Z_i' = Z_i - Z$, where Z stands for the mean of all Zs and Z_i' represents the estimated value at location i. The first order drift is denoted as $Z^\wedge_i = a + bZi$ and $Z_i' = Z_i - Z^\wedge$, where $Z^\wedge_i$ represents the estimated drift at point i. Likewise, the second order drift is denoted as $Z^\wedge_i = a + b Z_i + cZ_i^2$ and $Z_i' = Z_i - Z^\wedge$. In general, kriging first removes the drift prior to estimation, then estimates the semivariogram based on residuals. Due to its computational complexity, the kriging estimation process is much more time-consuming than other methods of spatial interpolation.

Surface Analysis Applications

Spatial interpolation is useful for a variety of applications. In most cases, the procedure starts with a set of irregularly spaced data points of known z values. These data points can be converted into either a grid, through gridding, or a set of triangular facets, through triangulation. The interpolated values are then used for the following applications.

Isarithmic Mapping

The most fundamental application of the interpolated z values is isarithmic mapping, or generating isolines to represent the z variable distribution. An isoline is a line connecting points of the same value of the z variable on the surface. Isolines are named according to various z variables. For instance, contours are elevation isolines, isobars are barometric pressure isolines, isotherms are temperature isolines, and so on.

In digital cartography, several rules govern the construction of isolines.

● Every isoline is theoretically enclosed either by itself or by the edge of the map. There should be no open ends of any isoline.

● Isolines do not cross one another. In special cases, overlapping isolines do exist, but they should never cross.

● As regards interpolation, every segment of an isoline separates the adjacent locality into two divisions with respect to the z variable. On one side of the segment, the z value is greater than the value assigned to the isoline, while on the other side the value must be less than that of the isoline.

In reality, some of the above principles may be violated due to special conditions. For spatial analysis, however, these rules must be maintained in order for the surface value to be accurately interpreted.

Isarithmic mapping requires that the z variable be organized either in a grid or a TIN. The next illustration shows an example of contour construction from a grid. In this case, the grid consists of only nine data points with the z values ranging between 14 and 28. To construct the isoline of 20, one could start at the segment on the upper left, where the values at both ends are 18 and 23. Apparently, there must be a point along this segment where the value is 20, assuming that the surface is continuous.

A contour linking points A-B-C-D-E constructed from the gridded data points.

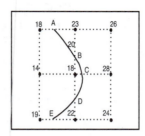

The position of the 20 point can be estimated between the two ends, depending on the difference in the z variable. In this case, it is located at point A which is two-fifths of the segment from the left end. At point A, the analyst searches for another point of the same value (20) from the near segments. Among the three possible segments, two are out of the question because their values at both ends are 14 and 18, and the value of 20 is out of range in both segments. The center segment has a range of 18 and 23; consequently, the location of the 20 value can be identified (point B). The search continues to connect points C, D, and E. The isoline can then be constructed by drawing straight line segments connecting the identified point A to E. A preferred and more aesthetic line can be drawn by connecting the points with a smoothing function resulting in the curve shown in the diagram.

When the z variable is organized in a grid, the search for the next point from an identified point always considers the three adjacent sides because there are four segments that define a grid cell. When the data are organized in TINs, then the search requires consideration of two edges of the same triangular facet. The following figure shows such an example. When point A is identified, the next point is to be identified from the other two edges of the same triangle. In this case, point B is identified because the values at the two ends of the other edge do not contain the value 20. The process continues on to identify the point C. Again, a smooth curve connecting the three points (A-B-C) of the same value (20) define part of the isoline.

A contour linking points A-B-C constructed from a TIN.

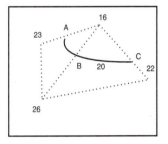

Isarithmic mapping is useful for depicting the distribution of any z variable. The following illustrations show the isohyets (precipitation) and isotherms (temperature) of the San Jacinto Ranger District in California.

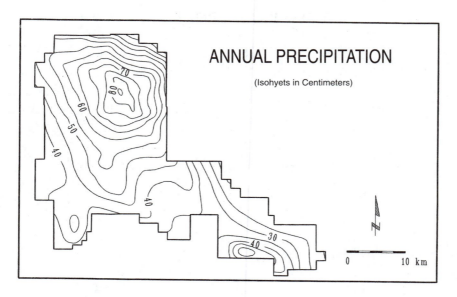

Isohyets (isolines of annual precipitation) in San Jacinto Ranger District, California.

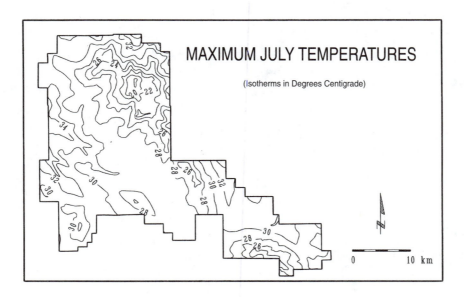

Isotherms (isolines of temperature) in San Jacinto Ranger District, California.

Topographic Profile

Topographic profiles, also known as *cross sectional profiles*, are derived from contour maps to show the variation in relief across a specific line segment or series of line segments. The figure at the left in the next illustration shows a typical contour map in which a topographic profile is to be constructed along the cross section P-Q. In this figure, the scale of the contour map is 1:2,400 and the contour unit is feet. Contour values vary from 40 to 80 ft. Conventionally, a contour with short hatches—in which the open ends of the hatches point to the lower surface—represents a depression. The contour map shows a valley on the left side and a depression under 80 ft at the right. To construct a topographic profile along the P-Q cross section, the vertical scale of the profile must be determined. Vertical scale is the ratio of the unit length (in actual distance) to the represented profile elevation.

Topographic profiles can be derived from contour maps.

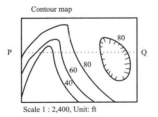

Contour map

Scale 1 : 2,400, Unit: ft

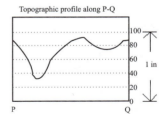

Topographic profile along P-Q

In the figure at the right, the actual length of the 100 ft elevation is 1 in. Thus, the vertical scale of the profile is 1 in: 100 ft or 1:1,200. Vertical scale is usually larger than the map (horizontal) scale so that the profile can exaggerate the variation in relief for better representation of the surface. Vertical exaggeration measures the degree to which the vertical scale is enlarged relative to the horizontal scale. In this example, the horizontal scale is 1:2,400 while the vertical scale is 1:1,200. Therefore, the vertical scale in the above figure—ratio of vertical to horizontal scale—is equal to 2. In other words, the profile represented at the right in the illustration exaggerates the vertical variation by a factor of 2.

The topographic profile along the P-Q section is thereby constructed by determining the elevation at intersections between the P-Q section and the contours in the figure at the left of the illustration. The elevations at those intersections are then copied to the corresponding positions in the figure at the right. As the identified points are connected, the curve becomes the topographic profile along the P-Q section.

In this example, the profile is constructed along a cross section which is a straight line segment. Profiles can also be constructed along several segments or linear features that are not straight lines. For instance, the profile along a road may show the actual gradient along the road, and thus the actual distance can be measured accordingly.

Perspective Diagram

Perspective diagrams, also known as block or fishnet diagrams, provide a 3D impression of a surface. Perspective diagrams are constructed by connecting topographic profiles in a certain order, such as along the x axis, along the y axis, or both.

The next illustration shows an example of a perspective diagram which is constructed from six topographic profiles along the x axis and six profiles along the y axis. Conventionally, when the diagram consists of profiles along both axes, it is called a *fishnet diagram*. The profiles are constructed first, and each is rotated and translated into a consistent system according to viewing angles (azimuth), viewing height (elevation), and viewing distance. Azimuth determines whether the diagram is being viewed from the south, north, northeast, and so on.

Perspective (fishnet) diagram of a surface.

Elevation determines how high the viewer is located relative to the surface. An elevation of 0 means that the surface is being viewed from the ground and an elevation of 90 means that the surface is being viewed directly from above. An elevation of 45 implies that the viewer is located at a 45-degree angle from the surface.

Viewing distance determines the structure of the perspective. If the distance is short, the viewer is close to the surface, and thus, the difference between the front (first) profile and the back (last) profile is great. In this case, the length of a distance on the front profile is much greater than the length of the same distance of the back profile. If the viewing distance is infinity, then there is no difference between the front and the back, and the same length of the same distance applies to front and back profiles. In addition to these three parameters, z scale or vertical exaggeration also affects the look of the perspective diagram in the variation of surface relief.

The following illustration shows a perspective diagram generated from the DEMs of El Cajon listed previously. The elevation (viewing height) is 30 degrees above the ground and the azimuth (viewing angle) is 45 degrees (from the southeast).

Perspective diagram: elevation is 30° and azimuth, 45°.

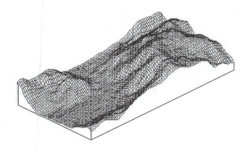

The next illustration shows the same perspective diagram viewed from a different angle. In this case, the azimuth is 135 degrees (from the southwest). The example demonstrates that the surface can be rotated in order to get a different perspective of the same area.

Perspective diagram: elevation is 30° and azimuth, 135°.

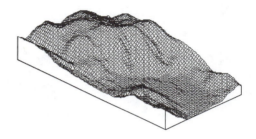

The viewing height can be altered to a different level. In the next diagram, the surface is viewed at a lower elevation. In this case, elevation is reduced to 15 degrees from the ground.

Perspective diagram: elevation is 15° and azimuth, 45°.

Vertical scale can be changed to either exaggerate or compress the relief. The next illustration shows the same perspective diagram with a z scale twice that of the preceding diagrams. When the relief is exaggerated, the difference in elevation is increased to make peaks look higher and valleys deeper.

Perspective diagram: z scale doubled.

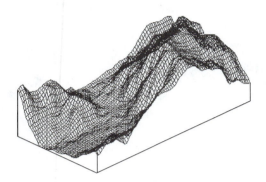

The above examples show how the same set of surface information can be manipulated in a GIS for analysis of the distribution. Furthermore, two perspective diagrams of different viewing positions can be carefully aligned to make a stereoscopic pair for true 3D viewing. Most people use a stereoscope to obtain such viewing results, while experienced analysts are able to look at stereoscopic pairs with the naked eye and maintain the 3D effects.

Thiessen Polygons

Thiessen polygons are polygons of irregular shape and size. An important property of Thiessen polygons is that they form equidistant boundaries between adjacent centroids. Given a set of irregularly spaced data points, Thiessen polygons can be constructed by first connecting each pair of adjacent points with a straight line segment, and then finding the midpoint along the segment and drawing a line vertical to the segment. This line forms the boundary between the two adjacent points. When all boundaries are found, a set of polygons is delineated. Assuming that the weight is identical for every point, Thiessen polygons show a system of territories delineated with equal influence between adjacent points. Although the surface characteristics are not homogeneous within the boundaries of a polygon, polygons delineated this way may be used to show areas dominated by the given data points.

Thiessen polygons constructed for three point features.

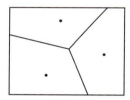

Other Applications

Surface analysis in GIS has a wide variety of applications in addition to the four discussed above. One useful feature of an interpolated surface is that information about the z value for any location in the map area can be effectively retrieved. The SPOT function in ARC/INFO allows the user to point at any location and retrieve the z value. This operation is straightforward because the surface value can be interpolated as long as the x and y coordinates of the location are known. This feature is useful because it can operate on any background coverage. For instance, a road map can be used as the background coverage and elevation can be identified for any spot from the road map, which is useful for planning a hiking trip, or for identifying a strategic location for military purposes.

Identification of visibility is another interesting application. In this case, the user may specify two point locations from the map in order to learn whether one spot is visible from the other. If two spots are not visible to one another, then the GIS also identifies the location where visibility is blocked.

Volume computation is a feature useful for construction and geological applications. In general, the volume between a hypothetical standard level (e.g., sea level) and a surface with known elevation data can be computed. For construction purposes, the required volume of cutting and filling can be estimated as long as two surface levels are specified. For geological applications, the volume of a geological structure can be computed by specifying elevations on the top of the structure and elevations on the bottom.

Another useful application is the detection of slope lines on a surface (Chou, 1992). A slope line is a line of steepest descent. Locations that determine a slope line follow directions that are vertical to contours. As such, slope lines can be detected from contours. Therefore, the most likely paths of debris flow after a landslide can be delineated if the site locations of potential landslides are specified. Furthermore, slope lines can be generated from a dense coverage of regularly spaced point locations, in which the generated lines represent the flow pattern of a hydrological structure. Locations where the lines merge are areas where slope failures are likely to occur and slope protection is required.

A recent study also applied surface analysis tools to correct surface area (Chou et al., 1995). Most GIS databases are obtained from 2D sources, such as aerial photographs or satellite imagery. The 2D representation of a surface distorts the true surface area in the sense that areas of steeper slope are reduced to a smaller surface area while areas of gentler slope are distorted to a lesser degree. With surface analysis capability, surface area can be correctly estimated provided that digital elevation models are available.

Summary

The distribution of any spatial phenomenon can be most effectively represented by a three-dimensional perspective diagram showing the surface of variations where peaks and valleys are easily identifiable. The quantity represented as the surface, commonly referred to as the z variable, can be organized in a GIS in different ways. When the z variable is organized in irregularly spaced points, the data volume is at its minimum yet the data are not readily available for either three-dimensional representation or surface analysis.

Surface information can also be organized in a series of digitized contours. Contours are two-dimensional representations of a three-dimensional surface. However, digitized contours are not suitable for surface analysis. In most GISs, surface information data are organized either in a grid/lattice format or as a set of triangulated irregular networks (TINs) for effective processing. With appropriate algorithms of spatial interpolation, contours can be derived from either a set of gridded data or TINs.

Spatial interpolation, the key process in surface analysis, involves the conversion of a set of discrete data points into a continuous distribution, based on the assumption that the change on the surface is gradual and consistent. The surface generated from different methods of interpolation could be quite different. A basic understanding of the available spatial interpolation methods is crucial for evaluating both the properties of a derived surface and the results of a surface analysis.

Possible applications of surface analysis are numerous. Cross-sectional profiles and perspective diagrams can be constructed for effective representation of the surface. Visibility from any site on the surface can be derived and the optimal location for a look-out tower identified. Cut-and-fill volumes of major construction projects can be estimated through the manipulation of multiple surface layers. Slope lines (lines of steepest descent on a slope) can be delineated for estimating the locations and volumes of debris deposit after a landslide. Surface area of polygon features on paper maps, aerial photos, or satellite imagery can be corrected using a GIS's surface analysis procedures.

Exercise

1. According to the point coverage of elevation data below, (a) construct a TIN using the shortest segment criterion, and (b) construct contours with contour values ranging from 10 to 20 at an interval of 5.

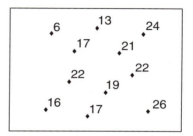

2. According to the contour map below, construct a topographic profile along the A-B cross section.

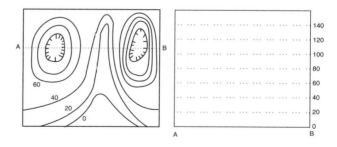

Grid Analysis

This chapter discusses operational procedures and quantitative methods for the analysis of spatial data in raster format. In grid analysis, geographic units are regularly spaced, and the location of each unit is referenced by row and column positions. Because geographic units are of equal size and identical shape, area adjustment of geographic units is unneccessary and spatial properties of geographic entities are relatively easy to trace. The regularity in the arrangement of geographic units allows for the underlying spatial relationships to be efficiently formulated. For instance, the distance between orthogonal neighbors (neighbors on the same row or column) is always a constant whereas the distance between two diagonal units can also be computed as a function of that constant. Therefore, the distance between any pair of units can be computed from differences in row and column positions. Furthermore, directional information is readily available for any pair of origin and destination cells as long as their positions in the grid are known.

Advantages of using the grid format in spatial analysis are listed below.

❏ *Efficient processing*. Because geographic units are regularly spaced with identical spatial properties, multiple layer operations can be processed very efficiently.

❏ *Numerous existing sources*. Grids are the common format for numerous sources of spatial information including satellite imagery, scanned aerial photos, and digital elevation models, among others. These data sources have been adopted in many GIS projects and have become the most common sources of major geographic databases.

❏ *Different feature types organized in the same layer.* For instance, the same grid may consist of point features, line features, and area features, as long as different features are assigned different values.

Grid format disadvantages appear below.

❏ *Data redundancy.* When data elements are organized in a regularly spaced system, there is a data point at the location of every grid cell, regardless of whether the data element is needed or not. Although several compression techniques are available, the advantages of gridded data are lost whenever the gridded data format is altered through compression. In most cases, the compressed data cannot be directly processed for analysis. Instead, the compressed raster data must first be decompressed in order to take advantage of spatial regularity.

❏ *Resolution confusion.* Gridded data give an unnatural look and unrealistic presentation unless the resolution is sufficiently high. Conversely, spatial resolution dictates spatial properties. For instance, some spatial statistics derived from a distribution may be different if spatial resolution varies, which is the result of the well-known scale problem.

❏ *Cell value assignment difficulties.* Different methods of cell value assignment may result in quite different spatial patterns.

The analytical methods discussed in preceding chapters assume a vector based, arc-node data structure. In principle, theories of spatial information and methods of spatial analysis are applicable to either data structure, raster or vector. The major difference between vector and raster based processing lies in the organization of geographic units, that is, polygons or grid cells. In GIS applications, however, these different data types are processed differently. The grid data operations and analytical methods presented in this chapter may not be directly applicable to vector based data. On the other hand, vector based polygons may be transformed into a grid structure and a grid data set may be transformed into a vector based polygon structure. However, resolution is sacrificed upon conversion in either direction.

Spatial Properties of Grid Data

A grid consists of data elements referenced by row and column. The number of rows may or may not be equal to the number of columns. Consequently, a grid may be a square or a rectangle. The referencing of rows and columns in a grid is different from the typical structure of the Cartesian coordinate system. In a coordinate system, the value of x along the horizontal axis increases to the right and the value of y along the vertical axis increases upward. The columns in a grid also increase to the right, yet the rows increase downward. The following figure illustrates the referencing structures of typical coordinate and grid systems. Although the values of x and y coordinates in a Cartesian system could be either positive or negative, all cells in a grid have a positive position reference, following the left-to-right and top-to-bottom data scan.

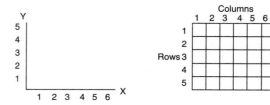

Every cell in a grid is an indivisible unit and must be assigned a value. Depending on the nature of the grid, the value assigned to a cell can be an integer or a floating point. When data values are not available for particular cells, the latter are described as "NODATA" cells. NODATA cells differ from cells containing a zero in that a zero value is considered to be data. NODATA cells are represented as blank cells in this book.

When a cell's vertical and horizontal sizes are equal, then the grid itself is considered a square. Conversely, when a cell's vertical and horizontal sizes differ, the grid is rectangular. Cell size represents the minimum geographic unit size and defines spatial resolution of the grid. Therefore, a high resolution grid contains a greater number of cells per geographic area than a grid of lower resolution.

Distance between grid cells can be computed in different ways depending on the definition of distance. For instance, in a square grid, the distance between two adjacent cells on the same row or column (i.e., orthogonal neighbors) can be equal to the size of the cell, which gives the centroid distance. To compute centroid distance between adjacent diagonal neighbors, use the row and column centroid distance multiplied by the square root of 2. For rectangular grids, the distance must be computed according to the height and width of each cell. Once the distance between adjacent cells is identified, the distance between any pair of cells can be computed from their positions referenced by columns and rows.

Data Value Assignment

The assignment of data values can follow different methods, each of which may result in a different grid definition. The three most widely adopted are the centroid, predominant type, and "most important type" methods. In addition, the hierarchical assignment method provides the most useful assignment solutions for spatial analysis.

Centroid Method

The centroid method assigns the value to a cell according to the observed value at its centroid location. In general, the centroid location of a cell is its geometric center. The next figure illustrates this method, in which a grid is overlaid on the original map. There are only two types of surface, indicated by shaded or blank areas. The shaded area could mean a specific land use type, a vegetation class, or the burned area, among other possibilities.

Value assignment based on the centroid method.

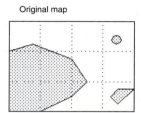

Original map

Assigned values

1	1	1	2
2	2	1	1
2	2	1	2

To translate the map into a grid, one of the two possible values, 1 or 2, is assigned to each cell depending on the observed value at the center of each cell. In this example, the cell at the lower right corner is assigned the value of 2 because its geometric center is in the shaded area, whereas the cell at the upper left hand corner is assigned a 1 because its center is in the blank area. This method is straightforward and is most effective when a large number of cells are of the same type with few local variations. A potential problem with this method is that the centroid may represent a minority type, such as the cell at the upper right corner. In this case, the assigned value may not be appropriate.

Predominant Type Method

This method assigns cell value based on the type of surface value with the largest area in the cell. The predominant type method involves more complicated computation than the centroid method because the area of every type must be computed for every cell. When the spatial resolution of a grid is low and the number of surface types is large, the assignment could be computationally complex. On the other hand, the assigned value tends to better represent the surface type of a cell. As seen in the next illustration, the cells at the upper and lower right corners are assigned a value of 1 instead of the value of 2 assigned by the centroid method.

Value assignment based on the predominant type method.

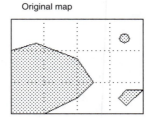

Original map   Assigned values

Most Important Type Method

The most important type assignment is another straightforward procedure in which the cell value depends on the occurrence of the most important surface type. This method is suitable when certain surface types are of critical concern in an analysis. For instance, a public health agency may need to report the location of contagious disease cases regardless of numbers per cell. The next figure shows the results of this method. In the example, the occurrence of the shaded type is considered most important. Any cell where this type of surface is observed is assigned the value of 2. The cells with a value of 1 indicate the total absence of the most important type.

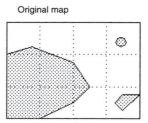

Original map

Assigned values

2	2	1	2
2	2	2	1
2	2	2	2

Value assignment based on the most important type method.

Hierarchical Method

Hierarchical assignment, the most complicated among available methods, requires that a decision tree be established before values are assigned to cells. The assignment of a cell follows a hierarchical structure of logical rules until a decision is reached. In general, results of this method are superior to other methods in terms of accuracy and usefulness. Because decision rules vary from case to case, they are programmed by the analyst.

Grid Operations

Common operations in grid analysis consist of the following functions: (1) local functions that work on every single cell, (2) focal functions that process the data of each cell based on the information of a specified neighborhood, (3) zonal functions that provide operations that work on each group of cells of identical values, and (4) global functions that work on a cell based on the data of the entire grid. The principal functionality of these operations is described in this section.

Local Functions

Local functions process a grid on a cell-by-cell basis, that is, each cell is processed based solely on its own values, without reference to the values of other cells. In other words, the output value is a function of the value or values of the cell being processed, regardless of the values of surrounding cells. For single layer operations, a typical example is changing the value of each cell by adding or multiplying a constant. In the following illustration, the input grid contains values ranging from 0 to 4. Blank cells represent NODATA cells. A simple local function multiplies every cell by a constant of 3. The results are shown in the output grid at the right. When there is no data for a cell, the corresponding cell of the output grid remains a blank.

A local function multiplies each cell in the input grid by 3 to produce the output grid.

Input grid

2	0	1	1
2	3	0	4
4		2	3
1	1		2

x 3 =

Output grid

6	0	3	3
6	9	0	12
12		6	9
3	3		6

Local functions can be applied to multiple layers represented by multiple grids of the same geographic area. The next illustration shows a typical case in which the input grid on the left is adjusted by a multiplier grid to produce the output grid on the right. For instance, the input grid shows the estimated value of each cell (parcel) based on the property value of buildings. The proximity to a transportation center from each cell is an adjustment factor; the values at the lower right corner must be adjusted higher because they are closer to transportation facilities. As such, the multiplier grid in the middle shows a pattern representing proximity to transportation at the lower left. The adjusted values are shown in the output grid. This example illustrates how local operations are applied to problems involving multiple data layers.

A local function multiplies the input grid by the multiplier grid to produce the output grid.

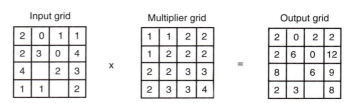

Local functions are not limited to arithmetic computations. Trigonometric, exponential, and logarithmic expressions are all acceptable for defining local functions. Logical operations can also be specified. For instance, the next illustration shows an example where the maximum value for each cell is to be identified from two grids. Grid 1 may represent the value of each parcel estimated by one appraiser while grid 2 represents parcel estimates by another appraiser. If the purpose is to determine the maximum estimated value for each cell, then the local operation of finding maximum values produces the output grid on the right. In this example, NODATA cells in at least one grid are treated as NODATA cells in the output grid, although this rule may be changed for other conditions.

*Local maximum
of the two input
grids produces
the output grid.*

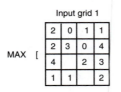

Focal Functions

   Focal functions process cell data depending on the values of neighboring cells. For instance, computing the sum of a specified neighborhood and assigning the sum to the corresponding cell of the output grid is the "focal sum" function. The following illustration shows an example of this focal function in which the neighborhood is defined by a 3X3 kernel. For cells closer to the edge where the regular kernel is not available, a reduced kernel is used and the sum is computed accordingly. For instance, the upper left corner cell is adjusted by a 2X2 kernel. Thus, the sum of the four values, 2, 0, 2 and 3 yields 7, which becomes the value of this cell in the output grid. The value of the second row, second column, is the sum of nine elements, 2, 0, 1, 2, 3, 0, 4, 2, and 2, and the sum equals 16.

*Focal sum function
sums the values
of the specified
neighborhood.*

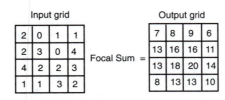

Another focal function is the mean of the specified neighborhood, the "focal mean" function. In the above example, this function yields the mean of the same kernel, that is, the eight adjacent cells and the center cell itself. This is the smoothing function to obtain the moving average in such a way that the value of each cell is changed into the average of the specified neighborhood. In the following illustration, for the cell at the upper left corner, the value in the output grid is the average of the four values 2, 0, 2, and 3, which approximately equals 1.8. Likewise, the value of the third row, second column in the output grid is the average of nine values (2, 3, 0, 4, 2, 2, 1, 1, and 3), which equals 2.0.

Focal mean function computes the moving average of the specified neighborhood.

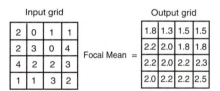

Other commonly employed focal functions include standard deviation (focal standard deviation), maximum (focal maximum), minimum (focal minimum), and range (focal range). In addition, the focal flow function identifies the flow direction for each cell based on the values of the adjacent cells, whereas the focal variety function identifies the diversity of values in the neighborhood.

The size and shape of the neighborhood can be defined using whichever method is appropriate for the analysis. The shape of the neighborhood can be a rectangle of any size, a circle, the annulus area between two concentric circles of different radii, or a portion of circle defined as a wedge.

Zonal Functions

Zonal functions process the data of a grid in such a way that cells of the same zone are analyzed as a group. A zone consists of a number of cells that may or may not be contiguous. A typical zonal function requires two grids—a zone grid which defines the size, shape, and location of each zone, and a value grid which is to be processed for analysis. In the zone grid, cells of the same zone are coded with the same value, while zones are assigned different zone values.

The next diagram illustrates an example of the zonal function. The objective of this function is to identify the zonal maximum is to be identified for each zone. In the input zone grid, there are only three zones with values ranging from 1 to 3. The zone with a value of 1 has five cells, three at the upper right corner and two at the lower left corner. The procedure involves finding the maximum value among these cells from the value grid. In this case, the maximum value is observed in the cell of the second row, fourth column and has a value of 8. In the output grid, all cells pertaining to this zone are assigned the maximum value of 8. Likewise, the zone with a value of 2 has five cells with a maximum of 5 in the value grid. Every cell in this zone is assigned the value 5 in the output grid.

Zonal maximum function identifies the maximum of each zone.

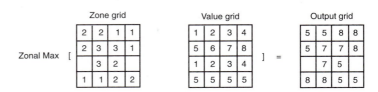

Typical zonal functions include zonal mean, zonal standard deviation, zonal sum, zonal minimum, zonal maximum, zonal range, and zonal variety. Other statistical and geometric properties may also be derived from additional zonal functions. For instance, the zonal perimeter function calculates the perimeter of each zone and assigns the returned value to each cell of the zone in the output grid.

Global Functions

For global functions, the output value of each cell is a function of the entire grid. As an example, the Euclidean distance function computes the distance from each cell to the nearest source cell, where source cells are defined in an input grid. In a square grid, the distance between two orthogonal neighbors is equal to the size of a cell, or, the distance between the centroid locations of adjacent cells. Likewise, the distance between two diagonal neighbors is equal to the cell size multiplied by the square root of 2. Distance between non-adjacent cells can be computed according to their row and column addresses.

In the next illustration, the grid at the left is the source grid in which two clusters of source cells exist. The source cells labeled 1 are the first cluster, and the cell labeled 2 is a single-cell source. The Euclidean distance from any source cell is always equal to 0. For any other cell, the output value is the distance from its nearest source cell. Thus, the cells adjacent to any source cell either in the same row or the same column have an output value of 1.0 which is the unit distance equivalent to cell size. Cells diagonal to a source have a value of 1.4. The cell at the upper left corner is 2 units from its nearest source cell, and thus has a value of 2.

Euclidean distance function computes the distance from the nearest source cell.

In the above example, the measurement of the distance from any cell must include the entire source grid; therefore this analytical procedure is a global function. Similarly, Euclidean direction identifies the direction of each cell to its nearest source cell. The function of Euclidean allocation determines which source cell each cell is closest to.

Global functions based on the Euclidean distance measurement assume that the distance between every pair of adjacent cells is a constant. In reality, the cost for traveling through two neighboring cells may be quite different from that for two other neighboring cells. In other words, the travel cost may vary from place to place depending on surface characteristics or transportation accessibility. The cost distance function is useful for adjusting the distance measurement by a cost grid.

The next figure provides an example of the cost distance function. The source grid is identical to that in the preceding illustration. However, this time a cost grid is employed to weigh travel cost. The value in each cell of the cost grid indicates the cost for traveling through that cell. Thus, the cost for traveling from the cell located in the first row, second column to its adjacent source cell to the right is half the cost of traveling through itself plus half the cost of traveling through the neighboring cell. Thus, the cost of the journey is 3. Likewise, the nearest source cell to the cell at the lower left corner is at the third row, second column, a diagonal neighbor. The travel cost is equal to half the cost of traveling through these two cells in a diagonal direction. Thus, the cost is 2.1.

Travel cost for each cell is derived from the distance to the nearest source cell weighted by a cost function.

Source grid

		1	1
			1
	2		

Cost grid

2	2	4	4
4	4	3	3
2	1	4	1
2	5	3	3

Output grid

5.0	3.0	0	0
3.5	2.5	2.8	0
1.5	0	2.5	2.0
2.1	3.0	2.8	4.0

Another useful global function is the cost path function, which identifies the least cost path from each selected cell to its nearest source cell in terms of cost distance. These global functions are particularly useful for evaluating the connectivity of a landscape and the proximity of a cell to any given entities.

Grid-based Spatial Analysis

Diffusion modeling and connectivity analysis can be effectively conducted from grid data. Grid analysis is suitable for these types of problems because of the grid's regular spatial configuration of geographic units. Diffusion modeling deals with the process underlying spatial distribution. The constant distance between adjacent units makes it possible to simulate the progression over geographic units at a consistent rate. Connectivity analysis evaluates interseparation distance, which is difficult to calculate in a polygon coverage, but can be obtained much more effectively in a grid.

Diffusion Modeling

A typical example of diffusion modeling is a fire behavior model in which the spatial progression of a fire is simulated. A probability field grid must be constructed first. Probabilistic spatial models, such as those generated through logistic regression (discussed in Chapter 8), are especially useful for deriving a probability field. In a probability field grid, each cell is assigned a fire occurrence probability value based on environmental and human-related variables identified as significant in wildfire distribution.

The following diagram shows a hypothetical grid of the fire risk probability field. The assigned probability for each cell is derived from a logistic regression model of fire occurrence probability. For the sake of simplicity, grid cells are classified into five categories. The dark cells represent areas of high fire danger where the fire occurrence probability is above 80%. The two clusters of high fire danger are surrounded by areas of fire occurrence probability of 40% to 60%. Blank cells, denoting areas of no fire danger, are fire breaks such as freeways, rivers, lakes, and the like. The linear features in the grid tend to represent either rivers or roads. The relatively large area of blank cells on the lower left corner of the grid may be a lake or a large parking lot.

Hypothetical grid of fire risk probability field.

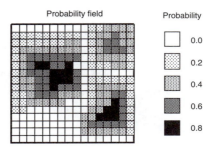

The probability field serves as the basis for simulating the progression of a fire. If ignition occurs in a specific cell, that cell is labeled as the initial burned cell. If information about local winds is available, then the probability field is adjusted by local winds. For instance, if local winds are northeasterly, then the probability for an adjacent cell located southwest of the initial burned cell to catch fire is increased, while the probability for northeast adjacent neighbors is adjusted lower. In addition, proximity to the initial burned cell must be taken into consideration when adjusting the probability field. Cells closer to the initial burned cell have a greater propensity to catch fire, while the probability for cells farther away is adjusted lower.

The grid below shows the distribution of adjusting factors calculated according to burned cells, proximity to burned cells, and winds. In this case, there is only one burned cell which represents the site of ignition. Adjustment of fire occurrence probability is applied to every cell with respect to the location of the ignition site. For the proximity adjustment, the distance to the ignition cell must be computed for every cell. Based on the computed distance, cells that are closer to the ignition site are assigned a higher adjusted value. In addition, if the direction of local winds is southwesterly, cells to the northeast of any burned cell receive a higher adjusted value while cells to the southwest receive lower adjusted values.

Grid showing distribution of adjusting factors.

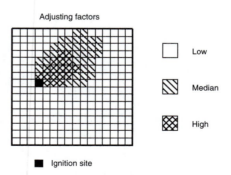

The adjusted probability field is converted into a grid in which each cell contains a range of integer numbers. The cells with a greater probability to burn have a wider range than cells with lower probability. A simulation procedure can then be designed in such a way that a cell is considered on fire if it is hit by a random number. In each iteration, a specified number of random numbers are generated; each generated number points to a specific cell. The burned cell becomes one of the source cells for the next iteration. The results of this procedure can later be compared with the actual diffusion process to verify the assumptions made in the probability assignment.

The diagram below illustrates the possible progression of a fire simulated through the grid analysis. In this example, the progression is simulated only to the third period. The first period (Time 1) contains only the burning ignition site. The fire will extend to the north and east during the second and third periods (Time 2 and Time 3). During the simulation process, the probability of a fire crossing a fire break of sufficient width, such as a lake, can be assumed to be zero. If a fire break is not wide enough, the fire may cross the barrier. The simulation procedure can continue by covering a longer time span. However, additional factors must be incorporated whenever necessary. For instance, suppression efforts tend to significantly alter the progression pattern.

Possible progression of a fire simulated through grid analysis.

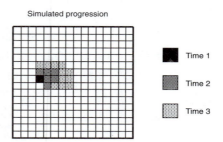

Simulated progression

■ Time 1

▓ Time 2

░ Time 3

Connectivity Analysis

The connectivity of a landscape measures the degree to which surface features of a certain type are connected. Landscape connectivity is an important concern in environmental management. In some cases, effective management of natural resources requires maximum connectivity of specific features. For instance, a sufficiently large area of dense forests must be well connected to provide a habitat for some endangered species to survive. In such cases, forest management policies must be set to maintain the highest possible level of connectivity.

On other occasions, connectivity of other feature types may have to be minimized. For instance, in wildfire management, areas of extremely high fire risk must be identified and the connectivity of such areas must be minimized in order to prevent fire holocausts. Quantitative methods must be developed for land managers to evaluate landscape connectivity and develop cost-effective spatial strategies to maintain a satisfactory level of connectivity.

Landscape connectivity analysis can be based on either vector based polygons or cell based grids. Although the theoretical principles are identical in both approaches, grid analysis possesses the advantage of constant spatial relationships among geographic units. The property of constant spatial relationships, which is especially useful for measuring connectivity among units, makes grid analysis an efficient tool for examining landscape connectivity.

The first step in connectivity analysis is deciding how connectivity should be defined and measured. In general, two geographic units are considered connected if the distance between them is within a specified threshold. Beyond the threshold, these units are not connected. In this regard, the measurement of distance must be based on interseparation instead of centroid locations.

In a typical grid based connectivity analysis, the landscape is divided into core, matrix, and barrier categories. All grid cells must be classified into these categories. For instance, cells dominated by certain types of forested lands at a sufficiently high density are classified as core cells. Core cells, for example, are essential for the survival of an endangered species.

Cells of other forest types or which are at a lower density are classified as matrix cells. In general, matrix cells are not suitable as habitat of the species in question, but instead provide an environment for the species to move around. In other words, if two core cells are connected by a matrix within the threshold distance, they are considered connected because the species can move from one core cell to the other. If a large number of matrix cells exist between two core cells, because the distance for traveling from one core cell to the other is too far for the species, then the distance between these two core cells exceeds the required threshold, and the cells are considered disconnected.

Barrier cells are the type of surface that prohibit any movement for the species in question. For instance, a freeway may be considered a barrier for deer to move from one side of the freeway to the other. Thus, if a barrier cell exists between two core cells, then these two core cells are considered disconnected. Classification of grid cells thus requires a local function in which every cell is processed and classified according to the nature of its surface.

Once all grid cells are classified into the three categories, the next step of the analysis is to identify the clustering of core cells. If two core cells are adjacent to each other, they must be considered parts of a single core unit instead of two separate core units. This step requires a focal function in which every core cell is evaluated based on its immediate neighbors to determine if it is a single, isolated core cell or part of a cluster of core cells. The step also requires the delineation of core clusters, such that core cells which are part of the same core cluster are merged to form a polygon-like core group. Operationally, each of such core cells is assigned the same code of core cluster, and thus, the congregation becomes a unique core zone which can be identified through zonal functions.

In addition, minimum size core zones can be specified, such that a core zone must contain at least a certain number of contiguous core cells. The minimum size consideration is necessary because the natural habitat of any species must be of a certain size; otherwise the species may not be able to survive. With the size consideration, the core cells that belong to any core cluster of a size less than the minimum required size are reclassified as matrix cells.

The next illustration shows the configuration of a landscape grid for connectivity analysis of forest fragmentation. In this example, the landscape is organized in a squared grid with 16 rows and columns. Each cell is a geographic unit of minimum spatial resolution.

Configuration of a landscape grid for connectivity analysis of forest fragmentation.

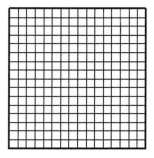

Forest density and maturity are considered in landscape classification. Forest density is evaluated by the number of grown trees per unit area, and maturity by average tree height within each unit area. In order to classify the landscape, the landscape grid must be overlaid with the forest density and tree maturity grid. Every cell in the overlaid grid is assessed by the classification criteria and the classified landscape is generated, as shown in the next diagram.

Classified landscape.

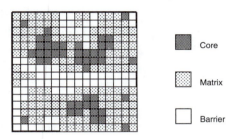

Core

Matrix

Barrier

In the classified landscape, cells that satisfy both density and maturity criteria are considered *core cells*. The core cells, which represent the natural habitat of a species, form three major clusters with additional single cells scattered about. *Matrix cells* satisfy either the density or maturity criterion, but not both. Such areas, although not suitable for providing natural habit, support the species' activity. The third landscape class, *barrier cells*, prohibit species movement, such as freeways, fences, structures, and so forth. Barrier cells divide the landscape into isolated activity spaces.

The classified landscape must be filtered by additional conditions. In this example, the commonly used size and edge filters are adopted. The *size filter* sets the minimum size for groups of core cells to be qualified as core clusters. This filter is necessary because forest coverage areas that satisfy both density and maturity criteria may not be large enough to provide a viable habitat for the species in question.

In this example, with the minimum size set at two grid units, single core cells are reclassified as matrix cells because they provide appropriate activity space for the species. The *edge filter* takes edge effects of the habitat into account. For instance, in order for a core cell to be part of the habitat, it must be protected by a certain distance from any barrier environment. In other words, all core cells exposed to barrier cells are not suitable for habitat. In the filtering process, core cells adjacent to barrier cells are reclassified into matrix cells. The filtered landscape, according to size and edge conditions, is depicted below.

Filtered landscape.

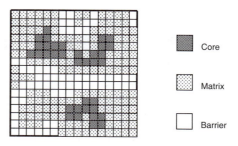

Core

Matrix

Barrier

By using global distance measurement functions, the interseparation distance between each pair of core clusters (unique core zones) can be determined. For this purpose, a cost function must be defined, and a local function must assign a cost value to every cell in the grid. The cost for traveling through a cell is zero for core cells, a unit distance for matrix cells, and a prohibitively high value for barrier cells. The cost path function can then be applied to calculate the minimum cost for each core zone to travel to every other core zone.

The distance is defined as the interseparation distance between core zones. If the distance is within the specified threshold, the two core zones are considered connected. Conceptually, if the interseparation distance between two core zones is too far, the core zones are disconnected and the species in question cannot move from one core zone to another through the connecting matrix cells. Core zones separated by barrier cells are always treated as disconnected because of the prohibitively high travel cost.

The following diagram illustrates the formation of core zones based on two logical conditions. First, core clusters linked through matrix cells with an interseparation distance (travel cost) less than a threshold size of two cell units are considered connected, and form a *core zone*. Core clusters separated by barrier cells are not connected and thus they form isolated core zones. Second, a matrix buffer equivalent to one unit is added to the core cluster to delineate the outer boundary of each core zone. In this example, two separate core zones are identified.

Formation of two separate core zones based on logical conditions forming a connected habitat.

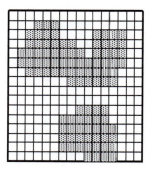

Once core zones are delineated, landscape connectivity can be derived according to a predefined measurement. Landscape connectivity can be measured in different ways. The simplest measurement, which evaluates connectivity mathematically, is the ratio of the number of core cells in qualified core zones to the total number of cells in the grid.

A more complicated measurement is based on the geometric relationships among the qualified core zones. Geometric connectivity is represented by the ratio of the number of connections between core zones to the total number of possible connections among such zones. In this case, a matrix of zonal connections must be established. The matrix has the same number of rows and columns as the number of core zones. Each cell in the matrix evaluates the inter-zonal connection between the corresponding pairs of core zones.

A third measurement of landscape connectivity is calculated from the topological relationships among core zones. Both direct and indirect connections between each pair of core zones must first be identified for this measurement. Two core zones may not be directly connected. However, if an indirect path can be established to connect them through one or more other core zones, they are considered indirectly connected.

The measurements of landscape connectivity described above may be further enhanced by adjusting each core cluster by its size (number of cells). In general, area adjusted measurements are more accurate in evaluating landscape connectivity.

Summary

Analysis of gridded data is an important area of GIS-based spatial analysis due to the existence of large volumes of geographic information organized in a grid format. An important advantage of grid analysis is that topological relationships among geographic units are embedded in the grid configuration, allowing spatial properties of the data to be effectively constructed. On the other hand, empirical applications of grid analysis require careful assessment of data redundancy and value assignment problems. The analyst must decide on trade-offs between spatial resolution and data volume to resolve data redundancy problems. Value assignment issues require that the analyst identify the most appropriate method for assigning values to grid cells.

For grid analysis, well-designed GISs such as ESRI's ARC/INFO provide efficient built-in utilities for vector-raster conversions as well as a wide range of operational functions. Such functions are typically classified into four categories depending on the spatial extent of each operation. Local functions act on a single cell at a time during each operation and the output is not affected by values of neighboring cells.

Focal functions apply to a clearly identified operational space that includes the cell being processed and its immediate neighboring cells. The shape and extent of the operational space can be defined according to the nature of the operation.

Zonal functions manipulate data in such a way that all grid cells are categorically classified into zones of homogenous space, and each zone can be processed differently. A major difference between a zone and the neighborhood space defined in focal functions is that cells belonging to a zone need not be contiguous. This property allows zonal functions to derive new data based on the properties of an existing grid without being constrained by the spatial configuration of the data.

Global functions operate on the entire frame when processing the gridded data. The output of each grid cell depends on the values of all grid cells and spatial distribution of the cells.

Most grid analyses employ some combination of local, focal, zonal, and global functions. Consistent topological relationships among grid cells and combined operational functions make grid based spatial analysis most useful for diffusion modeling and connectivity analysis. Diffusion modeling has a variety of possible applications, including wildfire management, disease vector tracking, migration studies, and innovation diffusion research, among others. Connectivity analysis is especially useful for natural resource and environmental management.

Exercise

1. *Value Assignment:* Construct a 16-by-16 square grid. Overlay the grid on the polygon map below and calculate the value of each cell based on the following assignment methods: (1) centroid, (2) predominant type, and (3) most important type. In the most important type method, importance is ranked by value where a lower value implies a higher level of importance. Compare the results of the three assignment methods.

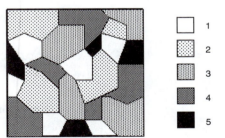

2. *Local Function:* Take the result of the centroid assignment method calculated in number 1 and convert the grid through a local function. The conversion adjusts the values of the grid to be

within the range between 0 and 1. The values in the original grid are converted in the following way: 5 in the original grid becomes 1 in the adjusted grid, 4 becomes 0.75, 3 becomes 0.50, 2 becomes 0.25, and 1 becomes 0. To carry out the conversion, you must specify a local conversion function and then apply this function to the original grid.

3. *Focal Function*: Smooth the grid created in previous exercises by applying a focal mean of a 3-by-3 kernel to the adjusted grid. Next, apply another local conversion function similar to the function described in number 2 to adjust the values to be within the range of 0 and 1 again. Compare the result with the outcome of number 2.

4. *Zonal Function*: Reclassify the grid created in number 3 into regular and target cells. Define the cells with a value of 0 as target cells without changing their values. Change the value of every non-zero cell to 1 and define it as a regular cell. The result is a grid with only two possible values, 0 and 1.

5. *Global Function*: Create a new grid showing the shortest Euclidean distance from each regular cell to the nearest target cell as defined in number 4.

Decision Making in Spatial Analysis

Chapters 1 through 10 presented general principles for manipulating and processing spatial information, and discussed standard spatial analysis methods using modern GIS technology. On some occasions, however, unusual constraints may exist and require specially designed solutions. Discipline-specific ana-lytical methods are available in textbooks on statis-tics, linear and non-linear programming, cartographic theory, transportation modeling, location-allocation modeling, survey engineering, remote sensing, graph theory, and topology, among others. Attempting to coordinate all such methods in a single volume is not only technically impossible but also practically unnecessary. When you face a complicated problem that requires more specific analytical methods, con-sult the references listed by chapter in the Reference section of this book.

Every analyst must select the most appropriate approach to a problem—qualitative or quantitative—and then the most appropriate method(s) within the chosen approach. In general, the qualitative approach is suited for projects in which variables are measured at the nominal or ordinal level. Quantita-tive analysis is generally most appropriate for projects based on variables measured at the interval or ratio level. While similarities exist between the two approaches, respective analytical procedures could be quite different. Depending on the nature of the problem and data availability, the analyst must decide which approach is more appropriate for the problem under consideration. The purpose of this chapter is to provide general guidelines for making decisions in spatial analysis design.

Chapters 1 through 5 presented principles, opera-tional procedures, and analytical methods that are common to both approaches, including single layer and multiple layer operations. Chapter 6, which focused on point pattern analysis, is useful in both approaches, although point patterns are less frequently used in a qualitative approach. Chapters 7 through 10 focused on the quantitative approach which solves specific types of research problems that require more diverse methods than a qualitative approach. It is also fair to say that the majority of spatial analyses using GISs employ quantitative data.

Certain variables or variable distributions cannot be evaluated quantitatively. For instance, it is difficult to assign a general soil property quantity for soils classified by type. In such cases, the qualitative approach is required for spatial analysis. However, because results of quantitative methods are more accurate, verifiable, and modifiable, the quantitative approach should be adopted whenever possible.

Qualitative Approach

Although most of the methods discussed in this book are quantitative in nature, GISs are also useful for conducting spatial analyses in a qualitative manner. In statistical terms, variables measured at the ordinal or nominal level are considered qualitative data. This section is partially derived from materials used in the Environmental Systems Research Institute's Planning Methods Seminar (Miller, 1995).

Definition of Issue

The first step in the qualitative approach is to define the issue. An issue is generally understood as what people are concerned about. When the concern can be expressed as a scaled value and the value can be represented on a map, the issue under consideration is a spatial issue. Geographic information systems deal with spatial issues where the involved entities are mappable.

Issues should not be confused with topics or data. A topic is a general area of understanding about the environment in question. For instance, earth science is a topic while seismic risk is an issue (area of concern) in earth science. Data are specific elements related to issues. Geology is a data category related to the issue of seismic risk.

In wildland fire management, the most important issue that people are concerned about is the distribution of fire risk. Related issues include the distribution of natural resources and the location of human properties. To study the issue of fire risk, the analyst must obtain data for several variables, including vegetation, climate, topography, existing preventive treatments, road networks, and so on. Each variable is to some degree related to the issue, but it is not the issue itself.

Once the issues related to the project have been clearly defined, they should be classified and the relationships among them identified. Operationally, project participants start with listing all potential issues and differentiate spatial from aspatial issues. Next, relationships among the listed issues must be established and the relative importance of each issue evaluated. Certain issues that may be incorporated into other, more general issues can be redefined. Issues considered relatively trivial can be eliminated from the list. The results of this step can be represented by a matrix which lists the related issues in descending order of relative importance. Each cell of the matrix is coded with the relationship between corresponding issues.

For example, in a spatial analysis aimed at developing spatial strategies for effective wildfire management, two main issues were identified: the distribution of fire occurrence probability and the distribution of related elements such as natural resources and human properties. Because both issues are spatial, their distributions can be mapped.

The next diagram shows a simplified decision matrix in which the cross impact between each pair of the four listed issues was evaluated. This matrix is considered simplified because empirical applications tend to involve more than four issues. The spatial issues identified for the task include land ownership (LO), fire risk (FR), concerned elements (CE), and management strategies (MS). These issues are all spatial and can be represented on maps. Based on the cross impact matrix, participants in the decision making process were asked to determine the relative importance of each pair in terms of their interrelationship. An "L" denotes a low level of relationship between the corresponding issues; "M" and "H" represent a medium and high level of the interrelationship, respectively.

Simplified decision matrix presenting cross impact of four issues.

	LO	FR	CE	MS
LO	-	L	L	M
FR	L	-	H	H
CE	L	H	-	H
MS	M	H	H	-

In reality, as more issues are involved in each project, the decision matrix becomes more complicated. This simplified example is used for illustrating how spatial issues can be evaluated in terms of interrelationships. The matrix indicates that land ownership is not closely related to either fire risk or concerned elements, although it is to some degree related to management strategies. The remaining pairs of spatial issues are highly correlated, that is, both fire risk and concerned elements must be considered when management strategies are evaluated and planned. Accordingly, land ownership can be eliminated from further consideration in wildfire management analysis.

Identification of Required Data

Once issues are clearly defined, the next step is to identify the data required for analysis of every issue under investigation. Any issue may be related to one or more variables, whereas the same variable may be related to one or more issues. Therefore, it is possible that the data for a particular variable will be shared by multiple issues. A diagram depicting the relationships between the set of variables and identified issues is useful for designing the structure of decision strategies in a spatial analysis.

When identifying required data, the analyst must examine both data availability and how the data for each variable are measured. If the data for a meaningful variable are not available, the analyst needs to assess whether collecting the data or finding substitutes would be more cost-effective. The analyst must also understand exactly how the data for each variable are measured because the measurement determines the way the variable should be used in the analysis.

In a GIS, the spatial distribution of each variable can be presented on a map. As such, the expected result of this step is a series of so-called data maps, along with well-organized metadata (data dictionary) explaining the definition and measurement of each variable.

In the example of wildfire management, potential variables related to fire risk include vegetation, slope gradient, slope aspect, proximity to roads, proximity to structures, temperature, and precipitation. Other variables of interest include locations of human structures, distribution of endangered species, and areas of important natural resources. For management tasks, roads, existing preventive treatments, resources available for suppression, and the like, are all relevant variables. Assuming that the data are available, a series of data maps can be constructed, with each showing the spatial distribution of each variable listed above.

The following diagram shows road data digitized in a GIS for analysis of wildfire management. Every line segment represents a section of existing roads with such attribute items as road type, slope, number of lanes, width, surface, and so forth. This data layer is useful for multiple purposes. Roads may be related to fire risk because human-caused fires tend to start at locations close to roads. Roads are important in fire management for both deployment of suppression forces and delineating the distribution of existing fire breaks.

Road data digitized in a GIS for analysis of wildfire management.

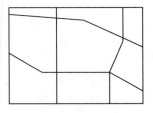

In the next diagram, dots represent human structure locations. Attribute items associated with each structure include structure type, size, material, and more. Structures are variables of interest in fire management because they require protection.

Dots represent human structure locations.

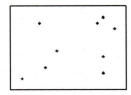

Original data may not be immediately available in a desirable format. For instance, the following map shows the triangulated irregular networks (TINs, discussed in Chapter 9) of the study area. This map is directly processed from raw data of the U.S. Geological Survey's DEMs. Each triangle represents a facet of homogenous topographic characteristics. In the attribute table, each triangular facet is coded with both slope gradient and slope aspect. Variables pertaining to elevation or topography can be obtained from this map.

TINs of study area, derived from U.S. Geological Survey DEMs.

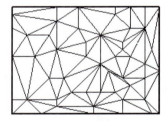

In a comprehensive ecological data set, the TIN map is not readily usable for spatial analysis, much less understandable to most users. With this original data map, topography can be interpreted through built-in GIS functions for surface analysis in order to generate two separate data maps: slope gradient and slope aspect. Slope gradient and aspect maps are more suitable than the original TINs in terms of both visual representation and spatial analysis.

The next diagram shows the slope gradient map processed from the TINs. This map is relevant to fire management because slope gradient affects fire behavior. In general, fires move faster on steeper slopes than on gentler slopes. For simplicity, triangular facets of similar slope gradient are classified into three categories appropriate for a qualitative analysis: low, gentle, and steep.

Slope gradient map processed from TINs.

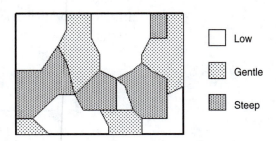

Likewise, triangular facets can also be processed to form polygons of similar aspect class. In the sample area, only four aspect classes are found on the surface: north (N), south (S), northeast (NE), and southwest (SW). Slope aspect is a valid concern for wildfire management due to the fact that fires behave differently on slopes of different aspect.

Slope aspect map.

⊞	S
☰	N
⊠	NE
⊠	SW

Moisture content and vegetation density vary over surface aspect. Moreover, large fires tend to occur during certain wind conditions. For instance, in southern California major fire outbreaks are associated with so-called Santa Ana conditions dominated by northerly or northeasterly winds. Clearly, the direction of local winds exercises a significant impact on fire behavior on different sides of a surface.

The above examples illustrate that original data maps are not readily usable for spatial analysis in most instances. Therefore, additional operational procedures are required for generating derivative maps suitable for analysis.

Derivative and Issue Maps

Once the data maps have been constructed, they are used to generate a set of derivative maps; the derivative maps are then used to generate issue maps. The derivative maps are useful for designing issue maps at the intermediate stage of analysis. Derivative maps may be derived either from a single data map or from a combination of data maps. In general, this step helps to reorganize the structure of the decision making process and reduce the problem to a smaller set of maps.

In the wildfire management example, the derivative vegetation map may be derived from the data map. In this case, the process to generate the derivative map involves simplification and generalization procedures. For instance, in a recent empirical study of fire distribution, the original vegetation map was coded into 23 vegetation species through aerial photo interpretation (Chou et al., 1990). It is not necessary to delineate vegetation polygons based on so many species in modeling fire occurrence probability. Therefore, a derivative map was produced by reclassifying the 23 species into general categories. These general categories included oaks, pine forests, high density chaparral, low density chaparral, and grass. In addition, the non-vegetated surface was divided into bare ground and water. Thus, although the original vegetation map contains 23 species plus non-vegetated surfaces, the derivative map contained only seven categories.

In addition to the vegetation map, a derivative map of topography was also generated. Conceptually, the map represents area units derived from two original data maps, one showing slope gradient distribution, and the other, slope aspect. Furthermore, for the elements related to fire management, a derivative map was generated from the maps of human properties, endangered species, and natural resources.

A set of derivative maps is used for determining the design of issue maps. Derivative maps generally differ from original source maps in that they are more suitable for detection of spatial patterns and are reclassified into categories more suitable for interpretation. Typically, polygons in each derivative map are rank ordered in terms of the specific subject of the spatial analysis. For instance, in the vegetation map, each general class of vegetation is assigned a code relating the vegetation class with fire propensity. In the topography map, each polygon may be assigned a code indicating the corresponding rate of fire spread.

Once derivative maps are generated, rules governing the decision making process must be established in order to translate the derivative maps into a set of issue maps. Each issue map depicts the distribution of a single identified issue. Procedures for translating derivative maps into issue maps are similar to those for preparing issue maps. The distribution of fire occurrence probability, for instance, can be derived from combining vegetation, topography, and climate maps, among others. In an issue map, each polygon is assigned a code indicating the level of propensity to burn. The issue map of fire-related elements is derived from the maps of human properties, endangered species, and important natural resources. Each polygon in this map is assigned a code indicating its relative importance. A polygon covering structures or habitat of an endangered species is classified as very important. A polygon located

at a remote site, away from any "concerned elements," is assigned a code of lesser importance.

As presented in previous examples, the slope gradient and aspect data maps are not readily usable in
the analysis. Therefore, a derivative topography map
is generated by overlaying the data maps and reclassifying the generated polygons. As depicted below,
the result is a derivative map showing surface units
of similar topography. The advantage of this map is
that each polygon represents a surface of homogenous topographic characteristics, in terms of both
slope gradient and aspect. Compared with the data
maps, this derivative map is more useful for evaluating properties pertaining to the surface structure.

*Derivative map
showing surface units
of similar topography.*

Likewise, the roads data map can be processed to
generate a more useful derivative map. In the following example, polygons can be created from buffer
zones of a specific distance from road segments to
represent areas that are close to roads. As human-
caused fires tend to start from locations close to
roads, this derivative map of road proximity provides

more useful information about the distribution of fire risk than the original data map.

Derivative of roads data map.

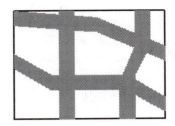

Another derivative map showing proximity to structures can be generated from the structures data map. As shown in the next illustration, buffer operations are again applied to point features of the original map to convert structure locations into areas of close proximity to structures. Such polygons represent one of the elements significant in fire management.

Buffer options applied to point features of original map, converting structure locations to areas near structures.

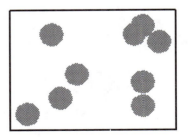

All derivative maps required for analysis must be constructed before issue maps can be generated. Issue maps are generated from derivative maps through the same operational procedures. In the example of wildfire management, an issue map showing fire risk distribution is crucial for both designing preventive spatial strategies and planning suppression efforts. Such a map can be produced through multiple layer operations on derivative maps of vegetation, topography, proximity to roads, and so on. The following illustration shows a typical fire risk issue map. For the sake of simplicity, the area is classified into three levels of fire risk. Polygons in this map represent areas where the probability of a fire breaking out is high, medium, or low. This issue map provides the basis for a variety of planning efforts in fire management.

Fire risk issue map.

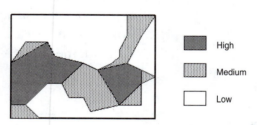

A second issue map required for wildfire management shows the distribution of concerned elements, or features of interest. This map can be obtained by overlaying derivative maps of the identified features, such as human structures, power lines, habitats of endangered species, important natural resources, and so forth. The following diagram shows a typical issue map of features of interest.

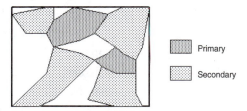

Issue map of features of interest.

Qualitative spatial analysis requires that maps depicting spatial distributions of issues relevant to the project be generated before the next step of morphological analysis. Morphological analysis evaluates relevant spatial issues based on the issue maps.

Morphological Analysis

Once all related issues have been assessed and the corresponding issue maps derived, the final step of the decision making process is to conduct a morphological analysis to provide solutions for the problem under consideration. In the qualitative approach, the results of a typical analysis tend to contain a number of feasible solutions instead of a single "optimal" solution. In most cases, the analysis helps to delineate a reduced set of feasible solutions. Each of the feasible solutions is associated with certain advantages and disadvantages. The morphological analysis is conducted to sort out the tradeoffs among feasible solutions and identify one or more most appropriate strategies. Strategies which fulfill study objectives by providing geographically centered solutions are called *spatial strategies.*

Morphological analysis is based on a matrix composed of issues. Each cell in the matrix indicates the assessed level of importance for the corresponding assignment. In the fire management example, assignment involves the allocation of management resources such as preventive treatments and suppression efforts. Areas that satisfy two conditions are delineated: areas of high fire risk indicated by a high level of fire occurrence probability, and areas where concerned elements such as human structures and endangered species exist. The delineated areas, representing critical zones for fire management, are rank ordered in terms of specific management needs. During the normal season, preventive treatments may be applied to these critical zones, depending on the availability of budget resources and the severity of the problem. When a fire breaks out, critical zones must be protected first while less important areas may require fewer suppression efforts.

The morphological analysis enables alternative management plans to be developed based on different sets of criteria. With a given budget, the cost-effectiveness of any management plan depends on the delineation of the most important objectives. However, each management plan should be assessed only according to a specific set of criteria. For instance, the hypothetical plan A below adopts the criterion of maximum protection. In this case, protection of features of interest (concerned elements) is of greatest importance. Overlapping zones of primary concern and highest fire risk require immediate clearing. In addition, preventive treatments will be allocated to areas where most features of interest are concentrated.

Hypothetical management plan.

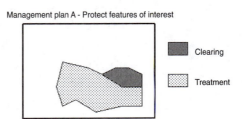

Management plan A - Protect features of interest

Clearing

Treatment

Alternatively, different criteria result in a different management plan. The next diagram illustrates an alternative plan B in which the top priority is fire risk reduction for the entire district. In other words, the principal objective of this plan is one of minimum fire risk. Areas of highest fire risk and primary interest require immediate attention. However, the allocation of preventive treatments will be concentrated in areas significantly threatened by fire risk.

Hypothetical management plan.

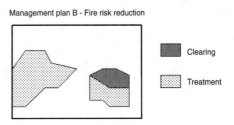

Management plan B - Fire risk reduction

Clearing

Treatment

In another example, assume that a land use plan is to be recommended to a planning agency. Ecological preserves, commercial centers, and residential areas are employed in the morphological analysis. Possible allocations of these three types of land use can be derived from an overlay of issue maps. Evaluation matrices are established to identify the most beneficial solutions. Alternative allocations are then evaluated in terms of how such land uses interact with one another in order to determine the most ideal solution.

This section described the general procedures of a qualitative approach. In reality, the concerns and technical problems for each project are unique. The analyst must consider both physical requirements and socioeconomic conditions. Decisions made under a qualitative approach tend to be influenced by subjective judgments and personal preference. Therefore, in most projects, decisions must be made in the context of an organized group of participants representing different interests and viewpoints.

Quantitative Approach

In general, the quantitative approach deals with variables measured at either the interval or ratio level. Most GIS operational procedures and associated analytical methods involved in quantitative spatial analyses were discussed in previous chapters. This section summarizes the decision making process in quantitative spatial analysis.

Problem Setting

The first steps are to clearly define the problem to analyze and define the study phenomenon. Depending on the nature of the problem and data availability, the study phenomenon may be represented as point, line, or area features.

Spatial phenomena represented as point features may be evaluated in different ways according to the nature of the problem. In the simplest case, each point represents a single occurrence of the study phenomenon. For instance, in a crime analysis, each dot represents the location of a reported incident. In this case, spatial analysis can be conducted based on the distribution of the point features or on aggregated frequency of point features within a set of predefined area units. When a study is based on the distribution of point features, the methods described in the "Point Pattern Analysis" of Chapter 6 are appropriate for the spatial analysis.

Alternatively, when point features are aggregated into groups, the frequency count in each geographic unit becomes the dependent variable. If the point features represent samples of a continuous surface, and each point is recorded with an attribute value such as elevation or crime rate, then spatial interpolation is required and the phenomenon is better represented by a surface. In this case, the observed points are translated into a set of triangulated irregular networks for spatial interpolation. Methods described in Chapter 9 should be employed.

Analysis of line features is based on the topology or density of line segments. If topology is the main concern, then network analysis methods as discussed in Chapter 7 are suitable. Alternatively, the density of line segments can be converted into a measure for area units.

Analysis of area features is most common in quantitatively oriented spatial analyses. Indeed, almost all significant spatial features can be treated as area features. For instance, natural resource plans deal with polygons of homogeneous environmental conditions; wildfire management strategies are based on management units of certain size; marketing research requires area delimitation based on socioeconomic characteristics; transportation planning applies to traffic analysis zones; and crime analysis can be formulated by city, zip code, census tract, and so on. Spatial modeling methods presented in Chapter 8 are appropriate for analysis of area features. The following discussion summarizes quantitative procedures for area feature analysis.

Definition of Geographic Units

Area units for spatial analysis can be defined in different ways. The grid and polygon methods are common. In the grid method, geographic units are defined by a grid of predetermined size and position. By overlaying the grid on a map of the study phenomenon, the analyst obtains a set of geographic units corresponding to the grid cells. This method is convenient for processing gridded data such as the

U.S. Geological Survey's digital elevation models (DEMs) or the digital data processed from satellite imagery. Furthermore, conducting spatial analyses using the grid method is much easier than using the polygon method due to the grid's homogeneity of geographic units. Topological relationships among geographic units can be systematically traced through row and column counts when using the grid method.

In spite of the above advantages, certain theoretical problems with the grid method should not be ignored. First, analysis results could be affected by the size and resolution of the grid. Next, as the spatial distribution of natural phenomena tends to be non-stationary, spatial statistics may be altered by placing the grid in a different way. Because the position of the grid is arbitrarily determined, the analyst has no way of knowing if a calculated coefficient is more valid than another derived from a grid that is placed differently.

The following diagram illustrates how grid placement affects the outcome of an analysis. The map at the far left shows a distribution of point features. Grids A and B are of identical size and spatial resolution and are placed on the point pattern in slightly different ways. The same system of value assignment is applied to both grids where a blank cell represents a unit of no occurrence of point features, a lightly shaded cell indicates a unit with a single point feature, and a dark cell contains two point features. Map pattern A shows the spatial pattern derived from grid

A while map pattern B is the result of grid B. This example demonstrates that the same distribution can create quite different spatial patterns according to the placement of the grid. Pattern A is a uniform or even distribution in which most cells were assigned the same value and most shaded cells are adjacent to shaded cells. Pattern B shows the opposite pattern where shaded cells are surrounded by blank cells. Estimated statistics of spatial autocorrelation will be quite different for these two patterns, although they are derived from the same original map.

Impact of grid placement on analysis outcome.

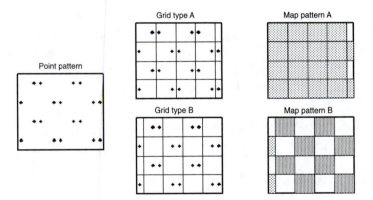

Alternatively, the polygon method defines geographic units according to criteria that are independent from the study phenomenon. Geographic units defined in this method tend to illustrate polygons of irregular shape and size. In addition, the distance between adjacent units is a variable rather than a constant. This method has been widely employed in studies that do not rely on gridded data. For example, in a study of a major measles epidemic in southwest England in 1969-70, maps based on local authority boundary areas revealed the spatial distribution of the spread of reported cases over time. A significant proportion of new reported cases was located adjacent to areas reporting cases the previous week (Haggett, 1972). Other examples include studies of international conflict in which geographic units are defined by political boundaries.

In typical disease vector studies, analyses of international conflict or trade, and socioeconomic or demographic models, geographic units must be defined by irregularly shaped polygons because they delimit the geographic distribution of data sources. For ecological and environmental applications, the polygon method is generally more appropriate than the grid method because polygons of homogeneous surfaces tend to reflect natural geographic units.

The polygon method requires one or more variables to be selected for defining geographic units. When selecting variables for this purpose, the questions appearing below must be considered.

1. *The polygons delineated from the selected variables must completely cover the entire study area.* If patches of unclassified surfaces exist, then spatial relationships among polygon features would be distorted.

2. *The delineated polygons themselves should not illustrate a clear spatial pattern.* Variables that generate a systematic distribution are not appropriate for defining geographic units. For instance, temperature zones delineated by isotherms that demonstrate a clear north-south pattern should

be avoided. In many cases, a variable measured at the nominal level is better than a variable measured at the interval or ratio scale in defining geographic units.

3. *The polygons delineated from the selected variables must define mutually exclusive geographic units.* For instance, polygons delineated by wildland fires may overlap because fires occur more than once in the same area. In this case, geographic units should not be defined by polygons representing wildland fires.

4. *The selection of variables for defining geographic units must also take into consideration the relationship between the selected variables and the study variable.* In general, a variable which is significant to the distribution of the study phenomenon is more desirable than an irrelevant variable. In studying the distribution of wildland fires, both vegetation and topography are reasonable candidates for defining geographic units because both are believed to be related to fire behavior. In other words, fires behave differently on slopes of different gradient or aspect and on surfaces of different vegetation cover. For such a study, census tracts are not a reasonable choice because these geographic divisions have no effect on wildland fires. However, census tracts may be an ideal choice for defining geographic units when a socioeconomic phenomenon is to be analyzed using census data.

5. *In order to avoid distortion of spatial relationships due to the varying size or distance among geographic units, variables that delineate poly-*

gons of relatively homogeneous shape and size are preferred. Vegetation in a desert environment is not appropriate for defining geographic units because its distribution tends to illustrate a spatial pattern where a number of extremely small islands of different vegetation types are surrounded by a vast non-vegetated area.

Polygons in the next illustration represent areas burned between 1911 and 1982 in the San Jacinto Ranger District, California. This layer is not suitable for defining geographic units for two reasons. First, areas that were burned more than once overlap. Second, fires vary significantly in size, and vast unburned areas exist.

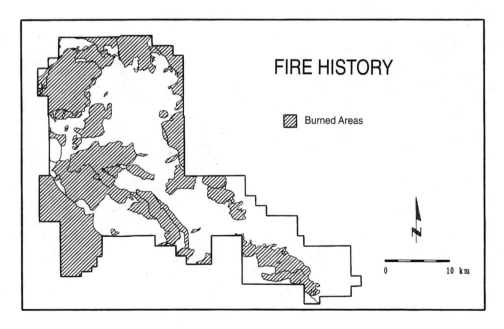

Fires in San Jacinto Ranger District, California, 1911-1982.

The following illustration shows polygons of homogeneous topographic characteristics derived from the triangular irregular networks of digital elevation models. Because topography is related to fire behavior and the fact that topographic polygons are similar in size and completely cover the study area, these polygons are suitable for defining geographic units in the study of wildfire distribution.

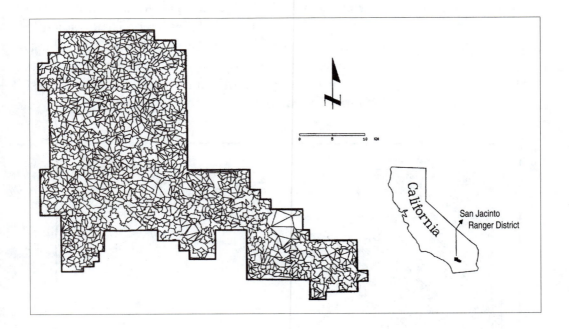

Polygons of homogeneous topographic characteristics derived from the triangulated irregular networks of digital elevation models.

Specification of Variables

When selecting a variable for inclusion in a spatial analysis, the analyst must consider the following three factors: significance of the variable, data availability, and measurement. A common problem in spatial analysis is that there are too many variables that may merit inclusion. On the other hand, data for critical variables may not be available.

In a comprehensive, ecological database, there may be many more variables than necessary for a particular analysis. For instance, in the abovementioned fire occurrence analysis, the database consisted of 17 data layers; several layers contained more than one variable. Another example would be a study of air transportation accessibility using census data. Census data contain numerous variables, and most of the variables are not useful for studying air transportation accessibility.

In some cases, it is not only impossible to incorporate all available variables in a model building process, but also unnecessary. The analyst must first evaluate the significance of each available variable and identify the list of variables to be incorporated. Variables that are neither meaningful nor significant in the distribution of the study phenomenon should be removed. An exploratory analysis can be conducted to determine the relationships among the available variables. Highly correlated variables (tested through correlation analysis) should be examined in order to identify a few key variables for analysis.

Data availability is a major concern in spatial analysis. More often than not, an existing database contains numerous trivial variables for the study at hand, while the data for key variables are not available. The analyst must determine whether other variables can be used as substitutes for key variables. Alternatively, the analyst must carefully assess how to obtain the required data and associated data collection costs.

Variable measurement is an important issue for variable specification. The analyst must understand exactly how every variable is measured. For instance, proximity to commercial centers may be measured by the straight line distance between a residence and the nearest commercial center, or by the street distance to travel from the residence to the nearest commercial center, or by the average distance from the house to all commercial centers in a given city. Alternatively, proximity to commercial centers may even be evaluated by an ordinal variable such as "far," "close," and so on. Measurement determines how results can be interpreted. The spatial resolution of the measurement and the scale of the original maps should be recorded in the data dictionary.

Construction of Statistical Models

When constructing statistical models, the analyst is concerned with the (1) dependent variable, (2) independent variables, (3) structure of the model, and (4) statistical testing procedure. Each element is discussed below.

Dependent Variable

The dependent variable represents the study phenomenon. For instance, in the abovementioned study of fire occurrence, fire distribution is the dependent variable. If the dependent variable is measured at a ratio scale, then a multiple regression model may be the first choice for testing. If the dependent variable is measured at a nominal scale, then logistic regression should be considered.

Independent Variables

The principal concerns here are selecting the variables and respective standards of measurement for incorporation in the model. Some variables may be highly correlated. Examination of the correlation between each pair of possible independent variables helps to identify their relationships. If the number of candidate variables is very large, a factor analysis or principal component analysis is useful for detecting the structure in the relationships among variables. Descriptive statistics should be generated for all candidate independent variables in order to identify low variance variables. The latter may not be appropriate for inclusion in the model.

Model Structure

Model structure should be considered once the independent variables are identified. The simplest structure is a simple linear additive model in which every variable takes an original form. Model structure can be modified by using the power function of a variable, the exponential function, or some combination of different forms. In the case of simple regression, the simplest model structure is expressed as follows:

$$Y = \beta_0 + \beta_1 X$$

The structure of the model may be altered into one of the following forms:

$$Y = \beta_0 + \beta_1 X^2$$

$$Y = \beta_0 + \beta_1 e^x$$

where e is the base of the natural logorithm

$$Y = \beta_0 + \beta_1 LOG(X)$$

Other model structures are available. A thorough exploratory analysis is necessary to determine the most appropriate form. For multiple regression models, the number of possible combinations of independent variables increases with the number of variables.

Statistical Testing Procedure

The determination of the most suitable structure can become a complicated task in spatial modeling. Different possible models must be tested for statistical significance. The structure which generates the highest level of significance is accepted as the most suitable model. In general, model structure and statistical testing deal with separate yet interrelated questions. To determine the most appropriate model structure, the analyst must examine the properties and descriptive statistics of both the dependent and independent variables, formulate possible structures based on prior knowledge about the relationships among variables, and conduct exploratory analyses such as correlation analysis, analysis of variance, and principal component analysis. In addition, when a model is constructed, examining the distribution of the error term vis-a-vis both dependent and independent variables could be helpful in identifying a more appropriate model structure.

The statistical testing procedure involves comparing models of different structures according to forecasting accuracy performance. For instance, in building different logistic regression models to explain bird distributions, the index of percentage correctly estimated (PCE) is used as an indicator of the maximum level of forecasting accuracy of each possible model. Model performance comparisons were made between different models in terms of the PCE index. For multiple regression models, the r^2 statistic is typically used to determine the extent to which the incorporated independent variables explain the variation of the dependent variable. In addition, the statistical significance of each estimated parameter must also be examined. For this purpose, the t-test and the χ^2 test of the significance of estimated parameters, discussed in Chapter 8, are useful for multiple regression models and logistic regression models, respectively.

Interpretation of Results

The final step in the quantitative approach is interpreting modeling results. The analyst must be aware of data limitations, variable measurement, limitations of the employed methods, and the level of statistical significance. With the use of a GIS, results can be displayed on maps which are especially useful for detecting the spatial pattern in the distribution of the phenomenon under investigation. Spatial relationships between the study phenomenon and independent variables can also be identified from maps. Interpretation of the results must be based on the spatial pattern of the distribution, while taking the limitations of the available data and employed methods into consideration.

For instance, the spatial pattern of bird distributions could be biased by the observation method. A spatial model of bird distribution may illustrate a clear relationship between the frequency of bird observations and surface type. The frequency of observation for areas of open space tends to be significantly higher than densely forested areas. The interpretation of the result could be misleading unless the correlation between the observation frequency and the difficulty of observation is taken into consideration. The lower observation frequency in densely forested areas may be associated with the difficulty of observation in such areas, while the higher frequency in open space may be a statistical bias toward easy detection.

When interpreting the results of a spatial analysis, the analyst must remember that spatial models do not imply causal relationships. For instance, crime incident distributions may be presented on a single data layer while fast food restaurants are represented as point features on another layer. The overlay of these two layers may illustrate a clear spatial correlation, that is, areas of higher density of crime incidents are associated with clusters of fast food restaurants. The spatial correlation does not suggest that fast food restaurants bring more crime incidents to respective neighborhoods, and neither does it imply that fast food restaurants were built in high crime-rate areas. In reality, both distributions may be affected by other socioeconomic or demographic characteristics. Consequently, the spatial correlation does not imply that the distribution of one phenomenon causes the other.

Summary

This chapter tackled the most fundamental questions of designing a GIS-based spatial analysis, that is, whether and how a qualitative or quantitative spatial analysis should be conducted. Principal concerns for each type of spatial analysis were summarized to provide a basis for choosing the most appropriate research tool.

GISs are primarily designed for quantitative spatial analyses because spatial information must be digitized and quantified in every database. Nevertheless, the use of GIS is not limited to quantitative analyses. In fact, common analytical functions that commercial GISs provide, including single layer and multiple layer operations, are also useful for solving qualitatively oriented problems. However, for problems that involve human perception, attitudes, personal feelings, subjective judgments, and the like, any simplified quantity would be, at best, misleading. For instance, how can one objectively measure the degree of happiness?

In GIS-based spatial analyses, a qualitative approach is often inevitable, especially when a research project requires input from numerous individuals of varying experience and perspective. At different stages of the decision making process, participants must be given the opportunity to assess related issues in such a way that personal opinions are represented. The analytical procedures outlined here include defining spatial issues, identifying required data for relevant and important issues, constructing derivative and issue maps, and conducting the final morphological analysis to reach a consensus.

Quantitatively oriented spatial analyses involve rigorous procedures of model building and validation, and thus are suitable for solving complicated spatial problems. Because every measurement must be objectively quantified, models built under this approach can always be enhanced by adding explanatory variables when they are shown to be significant, modified by refining model structures, and further improved by updating the estimated parameters when new data become available. Adequately defining a research problem, delineating geographic units, specifying dependent and independent variables, and constructing spatial models through rigorous statistical procedures are crucial to the application of modern GIS technology in a spatial analysis.

GISs are not only useful for mapping and database management. In fact, the most important use of GIS is spatial analysis. GISs provide very powerful analytical tools for a wide variety of disciplines that deal with spatial phenomena. Geographers, geologists, sociologists, environmental scientists, biologists, zoologists, entomologists, urban planners, transportation researchers, and many others can benefit substantially from recent advances in GIS technology.

Glossary

Arc-node data model

A vector-based data model designed for effective representation of line and polygon features. An arc represents a line feature defined by a start node and an end node, with any number of vertices in between. A polygon is defined by an enclosed set of connected arcs. An important advantage of this data model is the capability of tracing topological relationships among map features.

Area frequency

Total area of each class of polygon features classified by one or more attributes.

Area point

A point feature within an area feature which represents the area feature and carries associated attribute information. (See also **Entity point, Label point**.)

Area-adjusted mean

The arithmetic mean adjusted by polygon area, calculated by summation of the products of attribute value and polygon area, divided by total area. (See also **Arithmetic mean**.)

Arithmetic mean

The summation of all values of an attribute, divided by the number of cases.

Attribute data

Non-spatial data linked to layers. GIS software typically links spatial and attribute data to a single layer file. The attribute data linked to a particular layer may contain geographic information (e.g., addresses, zip codes, census statistics) or data associated with features in a layer such as soil properties or land use descriptions. A one-to-one relationship exists between features and attribute data records. Fields may be added to the attribute data linked to a layer to identify additional feature characteristics.

Attribute queries

Queries about attribute data without geographical referencing. Attribute queries do not require processing of spatial information.

Azimuthal projection

A projection of the graticule (parallels and meridians) from a sphere onto a tangent plane. On a map of azimuthal projection, great circle arcs with correct azimuths may be shown as straight lines for all directions from one or two points.

Bimodal distribution

A special distribution with two peaks (modes) on the curve of a frequency distribution.

Buffer operations

Construction of area features by extending outward from point, line, or polygon features over a specified distance.

Cartesian coordinate system

A coordinate system defined by two perpendicular axes with equal unit distance, usually referred to as the x axis and the y axis. In a Cartesian coordinate system, the location of a point feature is referenced by a pair of x and y coordinates. The system satisfies three spatial properties: (1) the distance from a point to itself is zero; (2) the distance from point A to point B equals the distance from point B to point A; and (3) the sum of the distance from point A to point B and the distance from point B to point C is greater than or equal to the distance from point A to point C.

Cartesian plane

A two-dimensional surface defined by a Cartesian coordinate system in which point features, line features, and polygon features can be expressed by a set of x and y coordinates.

Central tendency

A theoretical value of an attribute which represents the most likely value to be found in the distribution of that attribute. Depending on measurement scale, the central tendency of the attribute could be its mode, median, or mean.

Centroid method

A method of value assignment in grid analysis. The value assigned to a grid cell depends on the value identified at the centroid location of the cell, regardless of the rest of the cell. (See also **Predominant type method**, **Most important type method**, **Hierarchical method**.)

Complex polygon

A polygon with inner rings. Complex polygons may contain polygons. (See also **Polygon**, **Simple polygon**.)

Conformal projection

A map projection in which angular relationships among map features are preserved. The shape of every object is correctly represented although its area is distorted.

Conic projection

Projection of the graticule (parallels and meridians) from the sphere onto the surface of a cone, suitable for mid-latitude regions.

Connectivity analysis

Examination of landscape connectivity to determine the degree to which elements of a specific feature type are connected. For instance, if the habitat of an endangered species requires full-grown, dense forest, a connectivity analysis of such forest area will indicate whether a sufficiently connected activity space is available for the animals.

Constant interval classification

The classification of objects based on the value of an attribute with the same interval (range) for all classes.

Contingency table

A cross-classification table listing the frequency of observation for each combination of two attributes. For instance, if the attributes are soil and vegetation, then each entry in the table represents the frequency of a possible combination of a single soil type and a single vegetation type.

Convex hull

The outermost boundaries of a single polygon, or the combination of multiple polygons, that form a convex shape such that every straight line connecting any two points on the boundaries always lies inside the shape. In surface analysis, the outermost boundaries of a set of triangulated irregular networks always form a convex hull.

Coordinate system

Map reference system in which precise geographic position can be referenced for a locale by means of a rectangular grid. The use of a rectangular grid allows features to be located using x, y coordinates. Every coordinate system is derived from a specific map projection. (See also **Map projection**, **Universal Transverse Mercator (UTM)**, **State plane coordinate system**.)

Coverage

Refers to ESRI's ARC/INFO GIS. In ARC/INFO, a coverage is a database that stores geographic and tabular (attribute) data in a set of files. The data files are organized within a common directory.

Cross sectional profile

In surface analysis, a cross-sectional profile delineates the variation in the z variable (e.g., elevation) along a specified segment crossing the map area.

Cylindrical projection

Projection of the graticule (parallels and meridians) from the sphere onto the surface of a cylinder.

Decile range

The dispersion of an ordinal scaled attribute, measured by the difference between the rank order of the class at the top 10% and the bottom 10%.

DEM

See **Digital elevation model**.

Descriptive statistics

Basic statistics that describe the general properties of a data set in terms of certain attributes, including mode, mean, variance, maximum, minimum, range, and so on.

Developable surface

The surface of an object that can be converted into a two-dimensional plane (flat surface) without distorting the spatial relationships among features represented on the surface. The surface of a sphere is not developable because features on a sphere cannot be converted into a two-dimensional plane without distorting spatial relationships. Both cylinders and cones are developable because the surface of either type of object can be converted into a flat surface.

Diffusion modeling

Construction of spatial models for explaining, and probably predicting, the progression of a phenomenon. The study phenomenon can be anything which changes spatial extent over time, such as wildland fires, the spread of a disease, the diffusion of an innovation, the seasonal migration of birds, and so on. In GIS-based spatial analysis, diffusion modeling is most effectively handled through the grid analysis.

Digital elevation model

A set of digital topographic data, abbreviated as DEM, developed by the U.S. Geological Survey that covers most areas in the United States. The most commonly employed digital elevation models follow the 7.5-minute quadrangles of the topographic map series at a scale of 1:24,000. In this DEM set, elevation data are recorded in a raster mode at a spatial resolution of 30 m.

Digital line graph

A set of digital geographic data, also known as 100K DLG, developed by the U.S. Geological Survey that covers most areas in the United States. The data are vector based in the form of digitized lines at a scale of 1:100,000. DLGs are available for several categories of line features, including contours, roads, rivers, and so on.

Digitized contours

One of several possible forms of digital topographic data. Topography is represented by a set of line features connecting locations of identical elevation. (See also **Digital line graph**.)

DIME

See **Dual independence map encoding**.

Dispersion

The degree to which existing data values of an attribute deviate from central tendency. A high level of dispersion implies a more scattered distribution and a lower level of dispersion implies a more concentrated distribution about the central tendency. (See also **Central tendency**.)

DLG

See **Digital line graph**.

Dual independence map encoding

The vector based digital data of street maps of the United States, commonly known as the DIME files, developed by the U.S. Census Bureau for the 1970 census and updated for the 1980 census, to cover major metropolitan areas. Street maps are coded in two independent, yet interrelated, schemes—one contains line segments with information about nodes of street ends or intersections, and the other contains line segments with information about enclosed areas such as blocks, census tracts, cities, and so on. (See also **Geographic base file**.)

Easting

In the UTM coordinate system, easting is the measurement of distance in the east-west direction from the center meridian of a zone. The center meridian is arbitrarily assigned an easting of 500,000 m. An easting value greater than 500,000 m indicates a location to the east of the center meridian with the distance measured by the difference. An easting less than 500,000 m indicates a location to the west of the center meridian. (See also **Northing**.)

Edge

In network analysis, an edge represents a line segment connecting two point locations defined as nodes.

Entity point

A point feature representing the location of an object and carrying the attribute information associated with that object. An entity point is only meaningful in its position.

Equal-area projection

A map projection which preserves the area property so that all area features or regions on the map are shown in correct relative size.

Equal-distance zones

Zones generated from buffer operations to delineate areas within the same distance range from certain specified features. For instance, equal distance zones generated from freeways at an increment of 100 m show areas within the 100 m zone, 200 m zone, 300 m zone, and so on.

Euclidean distance

The straight line distance between two point locations represented on a two dimensional plane, measured by the square root of the sum of squared quantities of the difference between x coordinates and the difference between y coordinates.

Feature density

Density of features within a designated geographical area.

Fishnet diagram

Representation of a surface with a set of intersecting lines resembling a fishnet, also known as a *perspective diagram* or *3-D block diagram*. (See also **Perspective diagram**.)

F~mode~

F_{mode}

The frequency count of the mode in a distribution. The mode of a distribution is the value which occurs most frequently.

Focal functions

In grid analysis, a focal function processes the value of a grid cell based on the values of a designated neighborhood. The neighborhood can be any user-specified shape or size.

GBF

See **Geographic base file**.

Geographic base file

A digital geographic database developed by the U.S. Census Bureau for the 1980 census containing descriptive information of geographic areas integrated with the DIME files.

Geographic coordinates

Refers to the geographic reference system of latitude and longitude in which the Earth is treated as a sphere and divided into 360 equal parts (degrees). This division is performed along two axes, one running east-west along the equator, and the other running north-south along the Greenwich Prime Meridian. Using this coordinate system, any location on the Earth can be identified with a unique latitude-longitude pair. Geographic coordinates are commonly measured in degrees, minutes, and seconds, and can also be formatted as decimal degrees. (See also **Coordinate system**, **Map projection**.)

Geographic grid system

The worldwide accepted global coordinate system specified by latitude and longitude. Latitude measures the north-south angular distance from the equator. Longitude measures the east-west angular distance from the prime meridian passing through the Royal Observatory at Greenwich in England.

Geographic north

The orientation of a map defined by the direction pointing to the north pole.

Georeferenced data

Data geographically referenced by latitude and longitude, or by x and y coordinates of any other coordinate system that can be located on a map.

Global approximation

One of the two major approaches of spatial interpolation in surface analysis. Estimation of surface value at any point is based on the distribution of data values for the entire data set. (See also **Local estimation**.)

Global functions

In grid analysis, a global function derives the quantity of any grid cell based on the values of all cells in the entire grid. (See also **Focal functions**, **Zonal functions**, **Local functions**.)

Global positioning system

A system developed by the U.S. Department of Defense and designed for determining the precise position of any location on Earth. The system consists of 24 high-altitude satellites equipped with clocks of extremely high precision. There are at least four satellites visible to a receiver at any location on Earth. Precise position is calculated from the distance between the receiver and each visible satellite, while the distance is computed from the time difference between the satellite and receiver.

GPS

See **Global positioning system**.

Graticule

The imaginary network of parallels and meridians on which the geographic grid system for referencing locations on Earth by longitude and latitude is defined.

Grid analysis

A specific type of spatial analysis in which geographic space is represented by a regularly spaced, rectangle frame of a certain number of rows and columns. Spatial features and map objects are coded by assigning values to the corresponding grid cells while all analytical procedures are applied to one or more grid cells.

Grid cell

The indivisible geographic unit with a specifically assigned value in grid analysis, referenced by its corresponding row and column positions in a grid.

Grid north

Vertical line pointing to the northern end of a grid system. By convention, the vertical axis of a specific coordinate system follows the same direction as the grid north. The grid north may vary for different parts of the Earth and according to different coordinate systems. (See also **Magnetic north**, **True north**.)

Gridding

Construction of a grid from a set of irregularly placed data points using the estimated data value for each grid cell.

Hierarchical clustering

A mathematical procedure designed for effective classification of objects based on a specific attribute. The method starts with treating every object as a single object cluster. Each following iteration identifies one object and reassigns that object to another cluster, based on the criterion of minimization of within-cluster variance and maximization of between-cluster variance, until a suitable number of clusters is identified.

Hierarchical method

One of the available methods of value assignment in grid analysis. The assignment of attribute value to a grid cell follows a pre-established set of rules which specify the relative importance of attribute types and the procedures of assignment decisions. This method requires prior knowledge about the distribution of attribute values. (See also **Centroid method**, **Predominant type method**, **Most important type method**.)

Identity

Basic multilayer operation in which everything located within the boundaries of the input coverage (including information in the input and identity coverages) is collected in the output coverage when merging multiple data layers. In brief, the outer boundary of the output coverage is identical to that of the input coverage. All information from the identity coverage within the outer boundary of the input coverage is added to produce the output coverage.

Interseparation distance

The distance between two polygon features measured by the shortest possible distance that can be identified between respective boundaries.

Intersect

Basic multilayer operation equivalent to the Boolean AND operation in which two coverages are processed, but only the portion of the input coverage that falls inside the intersect coverage will remain in the output coverage. The additional information from the intersect coverage will be added to the output coverage as well. Everything in the output coverage is present in both input and intersect coverages.

Interval scale measurement

The measurement of an attribute which is meaningful in terms of the difference between any two values. For instance, temperature measured in degrees centigrade (Celsius) is an interval scale measurement because the same difference between any pair of values is equivalent (e.g., the difference between 5 degrees and 10 degrees is equivalent to the difference between 25 degrees and 30 degrees). However, an interval scale measurement is not meaningful in terms of ratio. In the case of degrees centigrade, 10 degrees is not twice as warm as 5 degrees. (See also **Nominal scale measurement**, **Ordinal scale measurement**, **Ratio scale measurement**.)

Isarithmic mapping

Construction of isolines based on a set of irregularly spaced data points. (See also **Isoline**.)

Isoline

A line formed by connecting point locations of identical attribute value. A *contour* is an isoline of elevation, which links locations of identical elevation. An *isobar* is an isoline of equal barometric pressure. An *isotherm* is an isoline of equal temperature.

Isotropic assumption

The assumption that the variation of data values is stable during rotation, that is, the spatial variation of data values is consistent in all directions.

Kriging

Theoretically, kriging is the most accurate method of spatial interpolation in surface analysis. In geostatistical terms, kriging is the optimal method of spatial linear interpolation, where the mean is estimated from the best linear unbiased estimator or best linear weighted moving average. The estimation of the data value at any location is based on the assumption of a continuous surface with variations consisting of the drift of regional tendency and the residual of local fluctuation.

Label point

A point position used for displaying text associated with a map feature. For instance, the label point for displaying the name of a polygon may be near its geometric center or anywhere inside the polygon.

Lambert conformal conic projection

One of the most commonly used conic projections suitable for mapping areas of east-west extent in the mid-latitudinal zones (e.g., the 48 conterminous states of the United States). The projection is conformal because it preserves the angular relationships among map features. This is one of the two projections employed for defining the state plane coordinate system. (See also **State plane coordinate system**.)

Latitude

See **Geographic coordinates**.

Line features

Geographic features represented on maps as one-dimensional objects, such as roads, rivers, or boundaries of administrative units.

Line segment

Direct line between two points.

Linear slope distribution

The frequency distribution of data values of an attribute can be described by a straight line of constant slope, implying a constant increase or decrease in frequency.

Link

In network analysis, a link denotes a line segment connecting two nodes. If a link does not exist between two nodes, the nodes are not directly connected.

Local estimation

In surface analysis, the value at a location can be interpolated from a set of nearby locations with known values. Estimation of the value at any location is based on the information of a specified set of adjacent locations rather than the entire set of data. (See also **Global approximation**.)

Local functions

In grid analysis, local functions are procedures that work on every grid cell independently of other cells. The derived value of a cell is not influenced by the values of surrounding cells. (See also **Focal functions**, **Zonal functions**, **Global functions**.)

Local search

In surface analysis, interpolation of the value at a location is based on local information of surrounding data points. Local search is the method to identify a specified number of data points to be included for interpolation.

Logistic regression

A special type of regression model with distribution of cumulative frequency best represented by a logistic curve (or an s-curve). In spatial modeling, logistic regression is suitable for situations when the dependent variable is of a dichotomous classification, such as yes or no, male or female, and the like, while independent variables are measured in an interval or ratio scale.

Longitude

See **Geographic coordinates**.

Magnetic north

Refers to the direction in which compass needles point. Magnetic north is different from true north and it varies over time depending on the magnetic field of the Earth. (See also **True north**, **Grid north**.)

Manhattan distance

Also known as *city block distance*. Distance between two locations is measured by the actual distance traveling on streets instead of straight line distance.

Map orientation

Direction of a map, typically pointing to north. A complete map orientation set includes true north, magnetic north, and grid north.

Map projection

System by which the curved surface of the Earth is represented on the flat surface of a map. The problem inherent in all map projections is preserving the properties of area, shape, distance, and direction present on the Earth's surface through transformation to a map surface. Because it is impossible to preserve all properties simultaneously, it is necessary to select a map projection optimized to preserve the most desired property. (See also **Azimuthal projection**, **Cylindrical projection**, **Conic projection**, **Pseudo-cylindrical projection**.)

Map resolution

The minimum size of any valid representation of geographic objects, or spatial features, on a map.

Map scale

The ratio of the unit distance on a map to the true distance on the ground that the unit represents. A 1:24,000 scale means that one inch on the map represents 24,000 inches (2,000 ft) on the ground.

Mean (average)

The sum of all values of an attribute divided by the number of cases.

Meridian

An arc which can be drawn on the surface of the Earth while satisfying the following properties: (1) it is exactly one half of a great circle on Earth; (2) it connects the north and south poles; and (3) it is perpendicular to the equator and any parallel. (See also **Parallel**.)

Metadata

Data about data (data dictionary) which describe the content and structure of a data set.

Minimal spanning tree

In network analysis, a minimal spanning tree is a specific network structure which has a designated node (a point location on the network) defined as the root of the structure and the network consists of shortest paths from the root to all other nodes on the network.

Mode

The attribute value of maximum occurrence for a nominal scaled attribute.

Moran's *I* correlation

The most commonly used measure of spatial autocorrelation, useful for describing the spatial pattern of a distribution. Value ranges between 1 and -1. A high value implies a clustered pattern, while a low value implies a scattered pattern. A value close to 0 indicates a random pattern.

Morphological analysis

The final step in a qualitative spatial analysis where maps showing all relevant and significant spatial issues are assessed with respect to the specified objectives of the analysis. Different criteria can be established for alternative spatial strategies to be derived. The final output of a morphological analysis includes a set of alternative management plans with explicitly expressed advantages and disadvantages associated with each plan.

Most important type method

In grid analysis, a value assignment method for assigning the value to a grid cell based on the most important feature (or value) present in the cell. A set of rules determining the relative importance of attribute values must be set prior to value assignment. (See also **Centroid method**, **Predominant type method**, **Hierarchical method**.)

Multiple cluster classification

Classification of geographic objects based on one or more attributes grounded on the assumption that the objects form multiple clusters of similar characteristics. In this case, the main tasks of a spatial analysis are to identify the optimal number of clusters and determine the members of each cluster.

Multiple layer operations

Also known as *vertical operations*, multiple layer operations are GIS procedures that apply to multiple data layers at the same time. Overlaying multiple layers to generate a combined data layer is a typical multiple layer operation. (See also **Single layer operations**.)

Multiple regression

Construction of a regression model for explaining the relationships between the dependent variable (the study variable) and two or more independent, explanatory variables.

National Committee for Digital Cartographic Data Standards (NCDCDS)

A committee organized by the American Congress on Surveying and Mapping (ACSM) in 1982 for the purpose of setting standards for digital cartographic data in order to facilitate the exchange and sharing of digital spatial information obtained by government agencies and private firms. The first set of standards were published in 1987 and later approved by the National Institute for Standards and Technology as a Federal Information Processing Standard.

Nearest neighbor index (NNI)

A measurement of spatial pattern in point pattern analysis based on the distance between every point feature and its nearest neighboring point. A small NNI value (close to 0) indicates a clustered pattern. A value close to 1 indicates a random pattern. A value significantly greater than 1 implies a scattered pattern.

Near

Fundamental proximity analysis function in a GIS which identifies the nearest point or line feature on one layer from a point feature on another layer, and computes distance accordingly.

Negative spatial autocorrelation

Spatial autocorrelation indicates the degree to which the distribution of a spatial phenomenon is influenced by the relationships among members of the phenomenon. Negative spatial autocorrelation implies that entities of the phenomenon tend to separate themselves in space and form a scattered distribution pattern. (See also **Positive spatial autocorrelation**.)

Neighborhood effects

The influence of surrounding conditions on the distribution of a spatial phenomenon. For instance, in modeling the distribution of birds, the analyst may need to consider how the selection of a nesting site is affected by the existence of other birds in the neighborhood.

Network accessibility

Network accessibility measures how easy it is to travel from a given node to any other node in a transportation network. A more accessible network is characterized by relatively larger numbers of choices and lower costs for movement between any origin and destination.

Network analysis

Spatial analysis of transportation networks represented as a set of nodes connected by line segments.

Network connectivity

Measures the degree to which nodes in a transportation network are connected through a set of line segments called *links*. Network connectivity depends on the number and structure of links.

Network diameter

In network analysis, the diameter of a network is the maximum of all the shortest paths between every node and every other node in the network, or the maximum extent of minimum internodal distances.

Nominal scale measurement

The categorical measurement of an attribute meaningful only by name. Nominal scale measurements are not meaningful in terms of rank order, difference, or ratio. Vegetation types are a nominal scale measurement because the name of a vegetation type does not carry any quantitative implications. (See also **Interval scale measurement**, **Ordinal scale measurement**, **Ratio scale measurement**.)

Normal distribution

One of several possible distributions of an attribute characterized by a bell-shaped frequency distribution that is symmetric on both sides of the curve's peak. Theoretically, the peak of the frequency distribution simultaneously represents the value of mode (value of maximum occurrence), median (value in the middle of all cases), and mean (the sum divided by the number of cases). Consult a standard statistical textbook for a more complete definition.

Northing

Measurement of distance in the north-south direction from the equator in the UTM coordinate system. (See also **Easting**.)

Ordinal scale measurement

Measurement of an attribute meaningful only in terms of rank order. For instance, the letter grading system (A, B, C, etc.) is an ordinal scale measurement because A is better than B, B is better than C, and so forth. Ordinal scale measurements are not meaningful in terms of the differences or ratio between values. For instance, the difference between A and B may not be equal to the difference between B and C, nor does A imply three times as good as C. (See **also Interval scale measurement**, **Nominal scale measurement**, **Ratio scale measurement**.)

Overlay analysis

Involves manipulation of spatial data organized in layers to create combined spatial features according to logical conditions specified in Boolean algebra.

Parallel

A circle drawn on the surface of the Earth that satisfies the following properties: (1) except for the equator, all parallels are small circles (i.e., smaller than the equator); (2) parallel to the equator and runs in the east-west direction; (3) intersects meridians at a right angle. (See also **Meridian**.)

PCE index

See **Percentage-correctly-estimated index**.

Pearson's correlation coefficient

Also known as *r*, the simple linear correlation coefficient. Pearson's correlation coefficient measures the degree to which two attributes are correlated to each other. Coefficient values range between +1 and -1. A value close to +1 indicates a high level of positive correlation, or an increase in one attribute implies an increase in the other. A value close to -1 indicates a negative correlation where an increase in one attribute implies a decrease in the other. A value close to 0 means a low level of correlation between two attributes.

Percentage-correctly-estimated index

The percentage of study cases that are correctly estimated by the calibration of a model when the parameters estimated in the model are employed to predict the distribution of the dependent variable.

Perspective diagram

Representation of a surface with a set of intersecting lines resembling a fishnet, also known as *fishnet diagram* or *3-D block diagram*. (See also **Fishnet diagram**.)

Pixel

A picture unit, which is also the minimum indivisible unit, in a raster database. The spatial resolution of a raster database is defined by pixel size.

Point distance

A proximity analysis function in a GIS which identifies all point features on one layer within a specified distance range from each point on another layer, and computes the distance between the points on respective layers.

Point pattern analysis

Analysis of spatial patterns illustrated in the distributions of geographic objects represented as point features on a map.

Point pattern

Spatial pattern of the distribution for a set of point features. The distribution may illustrate a scattered pattern, random pattern, clustered pattern, or other more complicated patterns.

Poisson process

In point pattern analysis, the Poisson process assumes that the frequency distribution of point features falling in any designated area follows a process of random permutation.

Polygon

A two-dimensional feature which occupies an area and consists of an outer ring and zero or more non-intersecting, non-nested inner rings. In ESRI's ARC/INFO, a polygon is delineated by an enclosed set of connected arcs (line features). (See also **Complex polygon**, **Simple polygon**.)

Polynomial trend surface method

Spatial interpolation method in surface analysis based on global approximation where the value at any point location is estimated from a polynomial function of its x and y coordinates.

Positive spatial autocorrelation

Spatial autocorrelation indicates the degree to which the distribution of a spatial phenomenon is influenced by the relationships among members of the phenomenon. Positive spatial autocorrelation implies that entities of the phenomenon tend to get closer together and form a clustered pattern in their distribution. (See also **Negative spatial autocorrelation**.)

Predominant type method

One of the available methods of value assignment in grid analysis. The attribute value assigned to a grid cell is based on the type of maximum frequency (maximum area of appearance) in the cell. (See also **Centroid method**, **Most important type method**, **Hierarchical method**.)

Prime meridian

Meridian running through the Royal Observatory at Greenwich near London, which divides the Earth into the eastern and western hemispheres. The prime meridian is the 0-degree longitude meridian and provides the basis for measuring longitude everywhere on Earth. The longitude of a location is the angular distance from the prime meridian. (See also **Meridian**.)

Proximal polygons

Also known as *Thiessen polygons*, polygons generated from a set of point features. Any point within a proximal polygon is closer to the point at the geometric center of the polygon than to any other point. (See also **Thiessen polygons**.)

Proximity zones

Zones of different distances from a specific set of geographic objects created through buffer operations in a GIS. For instance, a freeway proximity zone illustrates areas within a certain distance range from the freeway.

Pseudo-cylindrical projection

Projection of the graticule (parallels and meridians) from the sphere onto a developable surface. This projection resembles a cylinder and is based on mathematical methods aimed at preserving certain spatial properties.

Quadrat analysis

Analysis of a spatial phenomenon distribution by constructing a set of quadrats used to define geographic units. The quadrats can be of any size and shape, and may be placed randomly or regularly to form a grid on the study area.

Quantile classification

Classification system based on equal number of attribute values in each quantile. For instance, a 20% quantile classification indicates five classes. The top quantile would contain the highest 20% of the attribute values, the second highest quantile the next highest 20% of the attribute values, and so forth.

Ratio scale measurement

Meaningful measurement of a phenomenon in terms of the ratio between two different values. For instance, feet and kilometers are ratio scale measurements of distance (length/width): 8 ft is twice as long as 4 ft, and 100 km is 10 times as far as 10 km. (See also **Interval scale measurement**, **Nominal scale measurement**, **Ordinal scale measurement**.)

Relative location

The location of any geographic feature expressed as a relative measure from a referencing location. For instance, the location of a flag pole relative to a building can be expressed in relative terms such as direction and distance.

Remote sensing

Obtaining geographic information using a sensing device, such as camera or scanner, at a substantial distance from the target area. Typically, the sensing equipment is attached to aircraft or a satellite.

RF scale

Map scale expressed as a representative fraction such as 1:100,000 (1 to 100,000).

Saddle point problem

A common problem in spatial interpolation when two possible ways exist of interpreting the surface value of a location from four neighboring points.

Shortest path algorithm

In network analysis, the shortest path algorithm is a set of mathematical rules designed for identifying the path of shortest distance from one node to another.

Simple linear regression

The simplest regression model explaining the relationship between the dependent variable and a single independent variable. It is assumed that the relationship can be represented by a straight line.

Simple polygon

Polygon lacking inner rings. (See also **Polygon**, **Complex polygon**.)

Single layer operations

Also known as *horizontal operations*, single layer operations are GIS procedures applied to a single data layer at a time. Splitting a map layer into multiple blocks or appending (joining) multiple tiles to form a map of larger spatial extent are typical single layer operations. (See also **Multiple layer operations**.)

Sliver polygons

Little, unwanted polygons resulting from digitizing errors or processing spatial data from unmatched sources.

Spatial association

Spatial relationships between two sets of geographic objects. For instance, if landslide locations tend to be found on slopes of a specific soil type, a spatial association between landslides and soil polygons exists.

Spatial autocorrelation

Spatial autocorrelation indicates the degree to which the distribution of a spatial phenomenon is influenced by relationships among cases of the phenomenon. Positive spatial autocorrelation implies that cases tend to get closer together and form a clustered pattern. Negative spatial autocorrelation implies that cases tend to be separate in space and form a scattered pattern. (See also **Positive spatial autocorrelation**, **Negative spatial autocorrelation**.)

Spatial correlation analysis

Analysis of the correlation between two or more variables (attributes) based on geographically defined units. The correlation between soils and vegetation is spatial because the analysis is based on geographically defined units.

Spatial interpolation

Estimation of the value of an attribute, usually called the z variable, at any point location from the existing data of other locations.

Spatial modeling

Modeling relationships between different sets of geographic objects or attributes. A model is considered spatial if explicitly expressed spatial relationships are incorporated.

Spatial queries

Queries about attribute data with explicit geographical referencing. Processing of spatial information is required in spatial queries.

Standard deviation

Measurement of dispersion of an attribute about its mean, equivalent to the variance squared (sum of squared differences between each value and the mean, divided by the total number of cases).

State plane coordinate system

Coordinate system widely adopted for municipal mapping applications in the United States. The system is inconsistent over geographic areas. States of east-west extent are based on the UTM projection while states of north-south extent are based on the Lambert conformal conic projection.

String

An ordered sequence of line segments without nodes, node identifiers, or left and right identifiers.

Surface analysis

Spatial analysis of the distribution of an attribute, represented as the z variable, on the two-dimensional plane referenced by x and y coordinates.

Tessellation

Construction of a set of nested polygons to cover a study area. Different tessellation configurations include squares, triangles, hexagons, or polygons.

Thiessen polygons

Polygons generated from a set of point features, also known as *proximal polygons*. Anywhere within a proximal polygon is closer to the polygon's geometric center point than to any other point. (See also **Proximal polygons**.)

TIN

See **Triangulated irregular network**.

Topographic profile

Two-dimensional representation of a cross-sectional profile depicting the variation in surface relief along any designated line segment. (See also **Cross sectional profile**.)

Topography

Surface characteristics in terms of relief variation (distribution of elevation on the surface of the Earth).

Topology

Spatial relationships among spatial features. In GISs, the most important topological relationships include containment (polygons containing polygons), contiguity (polygons adjacent to polygons), and connectivity (arcs connected to arcs).

Trapezoid method

A trapezoid is a polygon with exactly four edges where two of the non-connecting edges are parallel to each other. The trapezoid method is employed in GISs for computing polygon area by dividing each polygon into as many trapezoids as the number of straight line segments that define the polygon, and adding up the areas of all trapezoids.

Triangulated irregular network

Organization of spatial information based on a set of triangles of irregular shape and size that form a connected network. Abbreviated as TIN, this structure efficiently organizes data of surface information with a minimum level of data redundancy.

Triangulation

Construction of triangulated irregular networks from a set of irregularly spaced data points.

True north

Direction pointing to the geographic north (north pole). (See also **Magnetic north**, **Grid north**.)

Union

Basic multilayer operation equivalent to the Boolean OR operation in which two or more data layers are overlaid to produce a combined coverage.

Universal Polar Stereographic coordinate system

Abbreviated as UPS, this system is complementary to the UTM coordinate system for polar areas beyond 84 degrees north or 80 degrees south. The system is based on a stereographic map projection centered at either pole.

Universal Transverse Mercator coordinate system

Abbreviated as UTM, this system is by far the most widely adopted in GIS applications in the United States. The surface of the Earth between the 84 degrees north latitude and the 80 degrees south latitude is equally divided into 60 zones, each covering a width of 6 degrees in longitude. Each zone defines an independent coordinate system from a transverse Mercator projection. Distance in the east-west direction is measured by easting, and north-south by northing. (See also **Easting, Northing**.)

UPS

See **Universal Polar Stereographic coordinate system**.

UTM

See **Universal Transverse Mercator coordinate system**.

UTM zones

In the UTM coordinate system, the area between 84 degrees north latitude and 80 degrees south latitude is divided into 60 zones of independent coordinate systems. Each zone covers a width of six degrees longitude. (See also **Universal Transverse Mercator coordinate system**.)

Valued graph

In network analysis, a valued graph is a transportation network consisting of nodes connected by a set of links where each link is assigned a value representing the impedance factor, such as travel time or travel cost.

Variation ratio

The measurement of dispersion of a nominal scaled attribute calculated as the ratio of the mode frequency to the total number of cases. (See also F_{mode}.)

Vertex/vertices

Specific x, y coordinate pair which makes up a line. Vertices are often referred to as *shape points*. The more vertices or shape points making up the line, the more accurate the representation of the feature.

Vertical operations

Simultaneous analytical procedures of geographic data manipulation applied on multiple data layers. Logical links connecting the spatial relationships among data layers are required in order to combine and process map features on different layers through vertical operations. (See also **Horizontal operations**.)

Z variable

Third dimension variable used in surface analysis. The x and y coordinates used for referencing geographic features are treated as the x and y variables that define a Cartesian plane. The z variable represents the attribute whose distribution is to be analyzed as a continuous surface. The z variable can be anything measured in an interval or ratio scale. For instance, elevation is the z variable for topographic analysis.

Zonal functions

In grid analysis, operational procedures that apply to cells of the same category according to a particular classification scheme. (See also **Local functions**, **focal functions**, **global functions**.)

Zone

In grid analysis, a zone consists of one or more grid cells belonging to the same category based on a specific set of classification criteria. Grid cells of the same zone may or may not be contiguous. For instance, when a vegetation grid is classified by major vegetation types, the grid cells of each type form a vegetation zone.

References

Chapter 1

Antenucci, J.C., K. Brown, P.L. Croswell, M.J. Kevany, H. Archer. 1991. *Geographic Information Systems, A Guide to the Technology*. New York: Van Nostrand Reinhold.

Aronoff, S. 1989. *Geographic Information Systems: A Management Perspective*. Ottawa: Canada: WDL Publications.

Burrough, P.A. 1986. *Principles of Geographic Information Systems for Land Resources Assessment*. Oxford: Clarendon Press.

Calkins, H.W. and R.F. Tomlinson. 1977. *Geographic Information Systems: Methods and Equipment for Land Use Planning*. Washington, DC: U.S. Government Printing Office, Department of the Interior.

Chou, Y.H. 1992. "Management of wildfires with a geographical information system." *International Journal of Geographic Information Systems*, 6:123-140.

Chou, Y.H., R.A. Minnich, and R.A. Chase. 1993. "Mapping probability of fire occurrence in the San Jacinto Mountains, California." *Environmental Management*, 17(1):129-140.

Chou, Y.H. 1993. "Nodal accessibility of air transportation in the United States, 1985-1989." *Transportation Planning and Technology*, 17:25-37.

Chou, Y.H. 1993. "Airline deregulation and nodal accessibility." *Journal of Transport Geography*, 1:36-46.

Chou, Y.H. 1995. "Automatic bus routing and passenger geocoding with a geographic information system." *Proceedings of International Conference on Vehicle Navigation and Information Systems*, Seattle, Washington, pp. 352-359.

Dent, B.D. 1990. *Cartography: Thematic Map Design*, 3rd edition. Dubuque, Iowa: Wm. C. Brown Publishers.

Environmental Systems Research Institute. 1994. *Understanding GIS: The ARC/INFO Method, Version 7 for UNIX and OpenVMS*. Redlands, California: ESRI.

Fotheringham, S. and P. Rogerson (eds.). 1994. *Spatial Analysis and GIS*. London: Taylor & Francis.

Laurini, R. and D. Thompson. 1992. *Fundamentals of Spatial Information Systems*. London: Academic Press.

Maguire, D.J., M.F. Goodchild, and D.W. Rhind (eds.). 1991. *Geographical Information Systems: Principles and Applications*. London: Longman.

Peuquet, D.J. and D.F. Marble. 1990. *Introductory Readings in Geographic Information Systems*. London: Taylor & Francis.

Robinson, A.H., J.L. Morrison, P.C. Muehrcke, A.J. Kimberling, and S.C. Guptill. 1995. *Elements of Cartography*, 6th edition. New York: John Wiley & Sons.

Star, J. and J. Estes. 1990. *Geographic Information Systems, An Introduction*. Englewood Cliffs, New Jersey: Prentice Hall.

Thomas. R.W. and R.J. Huggett. 1980. *Modeling in Geography, A Mathematical Approach*. Totowa, New Jersey: Barnes & Noble.

Unwin, D. 1981. *Introductory Spatial Analysis*. London: Methuen.

Wilford, J.N. 1981. *The Mapmakers*. New York: Vintage Books, Random House.

Chapter 2

Burrough, P.A. 1986. *Principles of Geographical Information Systems for Land Resources Assessment*. Oxford: Clarendon Press.

Chou, Y.H. 1992. "Slope-line detection in a vector-based GIS." *Photogrammetric Engineering & Remote Sensing*, 58:227-233.

Chou, Y.H., R.J. Dezzani, R.A. Minnich, and R.A. Chase. 1995. "Correction of surface area using digital elevation models." *Geographical Systems*, 2:131-151.

Clarke, K.C. 1995. *Analytical and Computer Cartography*. Englewood Cliffs, New Jersey: Prentice Hall.

Cromley, R.G. 1992. *Digital Cartography*. Englewood Cliffs, New Jersey: Prentice Hall.

Earickson, R.J. and J.M. Harlin. 1994. *Geographic Measurement and Quantitative Analysis*. New York: Macmillan College Publishing.

Environmental Systems Research Institute. 1994. *Understanding GIS, The ARC/INFO Method*. Redlands, California: ESRI.

Goodchild, M. and S. Gopal (eds.). 1989. *Accuracy of Spatial Databases.* London: Taylor & Francis.

Maguire, D.F. 1989. *Computers in Geography.* Essex, England: Longman.

Maling, D.H. 1989. *Measurements from Maps.* Oxford: Pergamon Press.

Masser, I. And M. Blakemore (eds.). 1991. *Handling Geographical Information.* London: Longman.

Monmonier, M.S. 1982. *Computer-Assisted Cartography, Principles and Prospects.* Englewood Cliffs, New Jersey: Prentice Hall.

National Committee for Digital Cartographic Data Standards. 1988. *The American Cartographer,* 15:1.

Star, J. and J. Estes. 1990. *Geographic Information Systems.* Englewood, New Jersey: Prentice Hall.

Tomlin, C.D. 1990. *Geographic Information Systems and Cartographic Modeling.* Englewood Cliffs, New Jersey: Prentice Hall.

Tomlinson, R.F., H.W. Calkins, and D.F. Marble. 1976. *Computer Handling of Geographical Data.* Switzerland: UNESCO.

Chapter 3

Bugayevskiy, L.M. and J.P. Snyder. 1995. *Map Projections: A Reference Manual.* London: Taylor & Francis.

Buttenfield, B.P. and R.B. McMaster (eds.). 1991. *Map Generalization.* Essex, England: Longman.

Campbell, J. 1991. *Map Use and Analysis.* Dubuque, Iowa: Wm. C. Brown.

Campbell, J. 1991. *Introductory Cartography.* Dubuque, Iowa: Wm. C. Brown.

Chou, Y.H. 1991. "Map resolution and spatial autocorrelation." *Geographical Analysis,* 23(3):228-246.

Clark, W.A.V. and P.L. Hosking. 1986. *Statistical Methods for Geographers.* New York: John Wiley & Sons.

Clarke, K.C. 1995. *Analytical and Computer Cartography,* 2nd edition. Englewood Cliffs, New Jersey: Prentice Hall.

Environmental Systems Research Institute. 1994. *Map Projections.* Redlands, California: ESRI.

Lewis, P. 1977. *Maps and Statistics*. New York: Wiley & Sons.

MacEachren, A.M. 1995. *How Maps Work: Representation, Visualization, and Design*. New York: The Guilford Press.

Maling, D.H. 1989. *Measurements from Maps*. Oxford: Pergamon.

McDonnell, P.W. 1979. *Introduction to Map Projections*. New York: Marcel Dekker.

Peterson, M.P. 1995. *Interactive and Animated Cartography*. Englewood Cliffs, New Jersey: Prentice Hall.

Richardus, P. and R.K. Adler. 1972. *Map Projections*. Amsterdam: North-Holland.

Robinson, A.H., J.L. Morrison, P.C. Muehrcke, A.J. Kimberling, and S.C. Guptill. 1995. *Elements of Cartography*, 6th edition. New York: Wiley & Sons.

Snyder, J.P. 1987. *Map Projections - A Working Manual*. USGS Professional Paper 1395. Washington, DC: US Government Printing Office.

Chapter 4

Aldenderfer, M.S. and R.K. Blashfield. 1984. *Cluster Analysis*. Beverly Hills, California: SAGE.

Dent, B.D. 1990. *Cartography Thematic Map Design*, 2nd edition. Dubuque, Iowa: Wm. C. Brown.

Environmental Systems Research Institute. 1990. *Introduction to Spatial Analysis: Spatial Manipulation and Analysis*. Redlands, California: ESRI.

Griffith, D.A. and C.G. Amrhein. 1991. *Statistical Analysis for Geographers*. Englewood Cliffs, New Jersey: Prentice Hall.

Robinson, A.H., R.D. Sale, J.L. Morrison, and P.C. Muehrcke. 1984. *Elements of Cartography*, 5th edition. New York: John Wiley & Sons.

Chapter 5

Clark, W.A.V. and P.L. Hosking. 1986. *Statistical Methods for Geographers*. New York: Wiley & Sons.

DeGroot, M.H. 1975. *Probability and Statistics*. Menlo Park, California: Addison-Wesley.

Dowdy, S. and S. Wearden. 1983. *Statistics for Research*. New York: Wiley & Sons.

Freund, J.E. 1971. *Mathematical Statistics*. Englewood Cliffs, New Jersey: Prentice Hall.

Haining, R. 1990. *Spatial Data Analysis in the Social and Environmental Sciences*. Cambridge: Cambridge University Press.

Henkel, R.E. 1976. *Tests of Significance*. Beverly Hills, California: SAGE.

Hogg, R.V. and E.A. Tanis. 1977. *Probability and Statistical Inference*. New York: Macmillan.

Theil, H. 1971. *Principles of Econometrics*. New York: Wiley & Sons.

Tufte, E.R. 1974. *Data Analysis for Politics and Policy*. Englewood Cliffs, New Jersey: Prentice Hall.

Yeates, M. 1974. *An Introduction to Quantitative Analysis in Human Geography*. New York: McGraw-Hill.

Wilson, A.G. and R.J. Bennett. 1985. *Mathematical Methods in Human Geography and Planning*. New York: Wiley & Sons.

Chapter 6

Boots, B.N. and A. Getis. 1988. *Point Pattern Analysis*. Beverly Hills, California: SAGE.

Chou, Y.H. 1991. "Map resolution and spatial autocorrelation." *Geographic Analysis*, 23:228-246.

Chou, Y.H. 1995. "Spatial pattern and spatial autocorrelation." *Lecture Notes in Computer Science*, 988:365-376. Springer-Verlag.

Clark, P.J. and F.C. Evans. 1954. "Distance to nearest neighbor as a measure of spatial relationships in populations." *Ecology*, 35:445-453.

Cliff, A.D. and J.K. Ord. 1981. *Spatial Processes: Models and Applications*. London: Pion Limited.

Cochran, W.G. 1977. *Sampling Techniques*, 3rd edition. New York: John Wiley & Sons.

Getis, A. and B.N. Boots. 1978. *Models of Spatial Processes*. London: Cambridge University Press.

Kershaw, K.A. *Quantitative and Dynamic Plant Ecology*. London: William Clowes & Sons.

Moran, P.A.P. 1948. "The interpretation of statistical maps." *Journal of the Royal Statistical Society*, Series B, 37:243-251.

Odland, J. 1988. *Spatial Autocorrelation*. Beverly Hills, California: SAGE.

Rohatqi, V.K. 1984. *Statistical Inference*. New York: Wiley & Sons.

Chapter 7

Bazaraa, M.S. and J.J. Jarvis. 1977. *Linear Programming and Network Flows*. New York: Wiley & Sons.

Bogart, K.P. 1983. *Introductory Combinatorics*. Boston: Pitman.

Chou, Y.H. 1990. "The hierarchical-hub model for airline networks." *Transportation Planning and Technology*, 14:243-258.

Chou, Y.H. 1993. "Nodal accessibility of air transportation in the United States, 1985-1989." *Transportation Planning and Technology*, 17:25-37.

Chou, Y.H. 1993. "Airline deregulation and nodal accessibility." *Journal of Transport Geography*. 1:36-46.

Chou, Y.H. 1993. "A method for measuring the spatial concentration of airline travel demand." *Transportation Research* B, 27(4):267-273.

Eliot Hurst, M.E. (Editor). 1974. *Transportation Geography: Comments and Readings*. New York: McGraw-Hill.

Environmental Systems Research Institute. 1991. *Network Analysis, ARC/INFO User's Guide 6.0*. Redlands, California: ESRI.

Ghosh, A. and G. Rushton. (Editors) 1987. *Spatial Analysis and Location-Allocation Models*. New York: Van Nostrand Reinhold.

Haggett, P. , A. Cliff, and A. Frey. 1977. *Locational Analysis in Human Geography*. New York: Wiley & Sons.

O'Kelly, M.E. 1987. "A quadratic integer program for the location of interacting hub facilities." *European Journal of Operational Research*, 32: 393-404.

O'Kelly, M.E. 1992. "Hub facility location with fixed costs." *Papers in Regional Science*, 71(3): 293-306.

O'Kelly, M.E. 1992. "A clustering approach to the planar hub location problem." *Annals of Operations Research*, 40: 339-353.

O'Kelly, M.E. and H.J. Miller. 1991. "Solution strategies for the single facility minimax hub location problem." *Papers in Regional Science*, 70(4): 367-80.

Taaffe, E.J. and H.L. Gauthier. 1973. *Geography of Transportation*. Englewood Cliffs, New Jersey: Prentice Hall.

Thisse, J. F. and H.G. Zoller (eds.). 1983. *Location Analysis of Public Facilities*. Amsterdam: North-Holland.

Wilson, A.G., J.D. Coelho, S.M. Macgill, and H.C.W.L. Williams. 1981. *Optimization in Locational and Transport Analysis*. New York: Wiley & Sons.

Chapter 8

Chou, Y.H., R.A. Minnich, L.A. Salazar, J.D. Power, and R.J. Dezzani. 1990. "Spatial autocorrelation of wildfire distribution in the Idyllwild quadrangle, San Jacinto Mountains, California." *Photogrammetric Engineering & Remote Sensing*, 56:1507-1513.

Chou, Y.H. 1992. "Management of wildfires with a geographical information system." *International Journal of Geographic Information Systems*, 6:123-140.

Chou, Y.H. 1993. "Airline deregulation and nodal accessibility." *Journal of Transport Geography*, 1:36-46.

Chou, Y.H. and S. Soret. 1996. "Neighborhood effects in bird distributions, Navarre, Spain." *Environmental Management,* 20(5): 675-687.

Cressie, N. 1991. *Statistics for Spatial Data*. New York: Wiley & Sons.

Dowdy, S. and S. Wearden. 1983. *Statistics for Research*. New York: Wiley & Sons.

Environmental Systems Research Institute. 1993. *INFO User's Manual*. Redlands, California: ESRI.

Griffith, D.A. and C.G. Amrhein. 1991. *Statistical Analysis for Geographers*. Englewood Cliffs, New Jersey: Prentice Hall.

Haining, R. 1990. *Spatial Data Analysis in the Social and Environmental Sciences*. Cambridge: Cambridge University Press.

Wilson, A.G. and R.J. Bennett. 1985. *Mathematical Methods in Human Geography and Planning*. New York: Wiley & Sons.

Chapter 9

Chou, Y.H. 1992. "Slope-line detection in a vector-based GIS." *Photogrammetric Engineering & Remote Sensing*, 58:227-233.

Chou, Y.H., R.J. Dezzani, R.A. Minnich, and R.A. Chase. 1995. "Correction of surface area using digital elevation models." *Geographical Systems*, 2:131-151.

Cressie, N. 1991. *Statistics for Spatial Data*. New York: Wiley & Sons.

Haining, R. 1990. *Spatial Data Analysis in the Social and Environmental Sciences.* Cambridge: Cambridge University Press.

Isaaks, E.H. and R.M. Srivastava. 1989. *An Introduction to Applied Geostatistics.* Oxford: Oxford University Press.

Sampson, R.J. 1975. "The surface II graphic system." In Davis, J.C. and M.C. McCullagh (eds.), *Display and Analysis of Spatial Data.* New York: Wiley and Sons. 244-266.

Chapter 10

Baudry, J., and Merriam, H.G. 1988. "Connectivity and connectedness: functional versus structural patterns and landscapes." In Karl-Friederich Schreiber (ed.), *Connectivity in Landscape Ecology.* Proceedings of the 2nd International Seminar of the "International Association for Landscape Ecology," Munster, 1987. 23-28.

Burrough, P.A. 1986. *Principles of Geograpical Information Systems for Land Resources Assessment.* Oxford: Clarendon Press. Chapter 5.

Chou, Y.H. and R.J. Dezzani. 1995. "Development of a model for evaluating landscape connectivity." Research Report 6CA38585, California Department of Forestry and Fire Protection, 1-277.

Environmental Systems Research Institute. 1991. *ARC/INFO User's Guide, 6.0, Cell-based Modeling with GRID.* Redlands, California: ESRI.

Environmental Systems Research Institute. 1992. *ARC/INFO User's Guide, 6.1, GRID Command Reference.* Redlands, California: ESRI.

Forman, R.T.T. and M. Gordon. 1986. *Landscape Ecology.* New York: Wiley & Sons.

Greenwood, G., and Eng, H. 1993. *Vegetation Projection and Analysis of the Cumulative Effects of Timber Harvest.* State of California Department of Forestry and Fire Protection. Strategic Planning Program.

Laurini, R. and D. Thompson. 1992. *Fundamentals of Spatial Information Systems.* London: Academic Press. Chapter 6.

Lillesand, T.M. and R.W. Kiefer. 1987. *Remote Sensing and Image Processing.* New York: Wiley & Sons.

Lombard, K. and R.L. Church. 1993. "The gateway shortest path problem: generating alternative routes for a corridor location problem." *Geographical Systems,* 1:25-45.

McDonnell, M.J., and Pickett, S.T. 1988. "Connectivity and the theory of landscape ecology." In Karl-Friederich Schreiber (ed.), *Connectivity in Landscape Ecology.* Proceedings of the 2nd International Seminar of the "International Association for Landscape Ecology." Munster, 1987. 17-22.

Philip, G.M. and D.F. Watson. 1982. "A precise method for determining contoured surfaces." *Australian Petroleum Exploration Association Journal*, 22: 205-212.

Tomlin, C. D. 1990. *Geographic Information Systems and Cartographic Modeling*. Englewood Cliffs, New Jersey: Prentice Hall.

Watson, D.F. and G.M. Philip. 1985. "A refinement of inverse distance weighted interpretation." *Geo-Processing*, 2: 315-327.

White, R. and G. Engelen. 1994. "Cellular dynamics and GIS: modelling spatial complexity." *Geographical Systems*, 1: 237-253.

Chapter 11

Burrough, P.A. 1986. *Principles of Geograpical Information Systems for Land Resources Assessment*. Oxford: Clarendon Press.

Chou, Y.H., R.A. Minnich, L.A. Salazar, J.D. Power, and R.J. Dezzani. 1990. "Spatial autocorrelation of wildfire distribution in the Idyllwild quadrangle, San Jacinto Mountains, California." *Photogrammetric Engineering and Remote Sensing*, 56:1507-1513.

Clark, W.A.V. and P.L. Hosking. 1986. *Statistical Methods for Geographers*. New York: Wiley & Sons.

Cliff, A.D. and J.K. Ord. 1981. *Spatial Processes: Models and Applications*. London: Pion.

Fotheringham, S. and P. Rogerson (eds.). 1994. *Spatial Analysis and GIS*. London: Taylor & Francis.

Laurini, R. and D. Thompson. 1992. *Fundamentals of Spatial Information Systems*. London: Academic Press.

Maguire, D.J., M.F. Goodchild, and D.W. Rhind (eds.). 1991. *Geographical Information Systems: Principles and Applications*. London: Longman.

Miller, W. R., 1995. "Planning Methods for GIS Environmental Design." Planning Methods Seminar, Environmental Systems Research Institute, Redlands, California.

Thomas. R.W. and R.J. Huggett. 1980. *Modeling in Geography, A Mathematical Approach*. Totowa, New Jersey: Barnes & Noble.

Unwin, D. 1981. *Introductory Spatial Analysis*. London: Methuen.

Yeates, M. 1974. *An Introduction to Quantitative Analysis in Human Geography*. New York: McGraw-Hill.

Index

More OnWord Press Titles

Computing/Business

Lotus Notes for Web Workgroups
$34.95

Mapping with Microsoft Office
$29.95 Includes Disk

Geographic Information Systems (GIS)

GIS: A Visual Approach
$39.95

The GIS Book, 3E
$34.95

INSIDE MapInfo Professional
$49.95 Includes CD-ROM

Raster Imagery in Geographic Information Systems
$59.95

INSIDE ArcView
$39.95 Includes CD-ROM

ArcView Exercise Book
$49.95 Includes CD-ROM

ArcView/Avenue Developer's Guide
$49.95

ArcView/Avenue Programmer's Reference
$49.95

101 ArcView/Avenue Scripts: The Disk
Disk $101.00

ARC/INFO Quick Reference
$24.95

MicroStation

INSIDE MicroStation 95, 4E
$39.95 Includes Disk

MicroStation 95 Exercise Book
$39.95 Includes Disk
Optional Instructor's Guide $14.95

MicroStation 95 Quick Reference
$24.95

MicroStation 95 Productivity Book
$49.95

Adventures in MicroStation 3D
$49.95 Includes CD-ROM

MicroStation for AutoCAD Users, 2E
$34.95

MicroStation Exercise Book 5.X
$34.95 Includes Disk
Optional Instructor's Guide $14.95

MicroStation Reference Guide 5.X
$18.95

Build Cell for 5.X
Software $69.95

101 MDL Commands (5.X and 95)
Executable Disk $101.00
Source Disks (6) $259.95

Pro/ENGINEER and Pro/JR.

Automating Design in Pro/ENGINEER with Pro/PROGRAM
$59.95 Includes CD-ROM

INSIDE Pro/ENGINEER, 3E
$49.95 Includes Disk

Pro/ENGINEER Exercise Book, 2E
$39.95 Includes Disk

Pro/ENGINEER Quick Reference, 2E
$24.95

Thinking Pro/ENGINEER
$49.95

Pro/ENGINEER Tips and Techniques
$59.95

INSIDE Pro/JR.
$49.95

Softdesk

INSIDE Softdesk Architectural
$49.95 Includes Disk

Softdesk Architecture 1 Certified Courseware
$34.95 Includes CD-ROM

Softdesk Architecture 2 Certified Courseware
$34.95 Includes CD-ROM

INSIDE Softdesk Civil
$49.95 Includes Disk

Softdesk Civil 1 Certified Courseware
$34.95 Includes CD-ROM

Softdesk Civil 2 Certified Courseware
$34.95 Includes CD-ROM

Interleaf

INSIDE Interleaf (v. 6)
$49.95 Includes Disk

Interleaf Quick Reference (v. 6)
$24.95

Interleaf Exercise Book (v. 5)
$39.95 Includes Disk

Interleaf Tips and Tricks (v. 5)
$49.95 Includes Disk

Adventurer's Guide to Interleaf LISP
$49.95 Includes Disk

Other CAD

*Manager's Guide to Computer-Aided
Engineering*
$49.95

Fallingwater in 3D Studio
$39.95   Includes Disk

Windows NT

Windows NT for the Technical Professional
$39.95

SunSoft Solaris

SunSoft Solaris 2. for Managers and
Administrators*
$34.95

SunSoft Solaris 2.* User's Guide
$29.95   Includes Disk

SunSoft Solaris 2.* Quick Reference
$18.95

Five Steps to SunSoft Solaris 2.*
$24.95   Includes Disk

SunSoft Solaris 2.* for Windows Users
$24.95

HP-UX

HP-UX User's Guide
$29.95

HP-UX Quick Reference
$18.95

Five Steps to HP-UX
$24.95   Includes Disk

OnWord Press Distribution

End Users/User Groups/Corporate Sales

OnWord Press books are available worldwide to end users, user groups, and corporate accounts from local booksellers or from Softstore/CADNEWS Bookstore: call 1-800-CADNEWS (1-800-223-6397) or 505-474-5120; fax 505-474-5020; write to SoftStore, Inc., 2530 Camino Entrada, Santa Fe, NM 87505-4835, USA or e-mail orders@hmp.com. SoftStore, Inc., is a High Mountain Press Company.

Wholesale, Including Overseas Distribution

High Mountain Press distributes OnWord Press books internationally. For terms call 1-800-4-ONWORD (1-800-466-9673) or 505-474-5130; fax to 505-474-5030; e-mail orders@hmp.com; or write to High Mountain Press, 2530 Camino Entrada, Santa Fe, NM 87505-4835, USA.

On the Internet: http://www.hmp.com

OnWord Press, 2530 Camino Entrada, Santa Fe, NM 87505-4835 USA